Achievement Motivation in the Leadership Role of Extension Agents

Achievement Motivation in the Leadership Role of Extension Agents

By

Roya Karami and Maimunah Ismail

CAMBRIDGE SCHOLARS PUBLISHING

Achievement Motivation in the Leadership Role of Extension Agents,
by Roya Karami and Maimunah Ismail

This book first published 2012

Cambridge Scholars Publishing

12 Back Chapman Street, Newcastle upon Tyne, NE6 2XX, UK

British Library Cataloguing in Publication Data
A catalogue record for this book is available from the British Library

ISBN (10): 1-4438-4150-1, ISBN (13): 978-1-4438-4150-4

We would like to dedicate our humble efforts in bringing out this significant publication for the benefit of all people particularly the scholars, professionals, public policy makers, our dear students, and all others who will use this book in their personal and professional lives in the area of Extension Education and beyond.

We also dedicate this book to:
Our parents, spouses, children, friends, and well-wishers who have encouraged us and provided support in many different ways, for making this effort a great success!

TABLE OF CONTENTS

List of Tables ix

List of Figures xi

Acknowledgements xiii

About the Authors xv

Introduction 1

Part I
Achievement Motivation in the Leadership Role:
A Theoretical Consideration

Chapter One 9
The Leadership Role in Extension Context

Chapter Two 31
Achievement Motivation in the Leadership Role

Chapter Three 53
Background Factors and Achievement Motivation in the Leadership Role

Chapter Four 69
Personal Factors and Achievement Motivation in the Leadership Role

Chapter Five 87
Environmental Factors and Achievement Motivation in the Leadership Role

Part II
A Study in Iran: Factors Influencing Achievement Motivation in Leadership Role of Extension Agents

Chapter Six 103
The Study: Methods, Respondents, and Organization

Chapter Seven.. 137
Achievement Motivation in the Leadership Role of Extension Agents: Evidence

Chapter Eight... 169
Achievement Motivation in the Leadership Role of Extension Agents: Future Agenda

LIST OF TABLES

6-1: Results of variables reliability in pilot test
6-2: The distribution of study clusters and questionnaires
6-3: GOF indices of the measurement model
6-4: The results of CFA for measurement model
7-1: Demographic and professional profile of respondents
7-2: Distribution of respondents by background factor variables
7-3: Distribution of respondents by personal factor variables
7-4: Distribution of respondents by environmental factor variables
7-5: Distribution of respondents by achievement motivation dimensions
7-6: GOF indices of direct structural model
7-7: Standardized regression weight in direct hypothesized path model
7-8: Results of independent- sample t-test on achievement motivation between male and female
7-9: GOF indices of overall structural model
7-10: Standardized regression weight in overall hypothesized path model
7-11: The effect of exogenous variables on achievement motivation through the mediation of personal and environmental factors

LIST OF FIGURES

2-1: The model of career motivation of London and Noe (1997)
2-2: Social cognitive model of Bandura (2001)
2-3: The achievement motivation model of Farmer's (1985, 1997)
2-4: The historical diagram of motivation theories and theoretical consideration of study
6-1: The research framework
6-2: Organizational chart of ministry of Jihad-e-Agriculture
6-3: Sampling procedure of the study
6-4: The measurement model of the study showing the relationships of items and constructs
7-1: The direct structural model of the study
7-2: The overall structural model of the study showing the relationships between the variables

ACKNOWLEDGEMENTS

I wish to avail myself of this opportunity, express a sense of gratitude and love to my beloved family: Mohammad Badsar my husband, Delvin and Narin my lovely daughters. I thank them for giving me moral and spiritual support. I would like to convey thanks to Universiti Putra Malaysia and Agricultural Research, Education and Extension Organization from Iran's Ministry of Jihad-e-Agriculture for my graduate work from 2008 to 20011. Words are inadequate in offering my thanks to Roohollah Reesi Sistani for providing editing services in this publication. I am also thankful to the respondents of the conducted study in Iran. I hope this conducted study provides basis to show the reliability and applicability of theoretical discussions presented in the first part of the book. My greatest appreciations to Prof. Dr. Maimunah Ismail for her tremendous support and help in this inspiring effort.

—Roya Karami

It is very rewarding to see the publication of this book based on a professional work of one of my dedicated students, Roya Karami. She has further proven to me the fact that the outcome of a graduate work is dependent on the working together between the supervisee and supervisor; and I have capitalised on the fact that 'graduate student is a motivated researcher'.

—Maimunah Ismail

ABOUT THE AUTHORS

Roya Karami
Email: rokarami@educ.upm.edu.my, rokarami@yahoo.com

Roya Karami is a post-doctoral fellow with a major in Extension Education at Universiti Putra Malaysia. She is also a member of the Agricultural Research, Education and Extension Organization from Iran's Ministry of Jihad-e-Agriculture. Her research areas include extension systems, career development, and achievement motivation. She has published articles in the areas of extension system and achievement motivation among extension agents in local and international journals.

Maimunah Ismail
Email: mismail@educ.upm.edu.my, mismail@ace.upm.edu.my

Maimunah Ismail, PhD, is Professor of Extension Education at Universiti Putra Malaysia. Her research areas include career development, corporate social responsibility, and gender studies. She has authored/co-edited 11 academic books and numerous articles in local and international journals. She sits on the Editorial Advisory Board of several international journals.

INTRODUCTION

This book is set out to illustrate factors influencing achievement motivation in the leadership role of extension agents. The book is structured in two parts consisting of eight chapters. The first part includes chapter one to five, and aims to theoretically discuss the achievement motivation in the leadership role of extension agents and its associated factors. The second part includes chapter six to eight, and aims to present an original study conducted in Iran using the theoretical considerations that are presented in the first part of the book. The second part of the book is established to show the reliability and applicability of constructs presented in the earlier chapters.

Chapter 1 introduces extension agents who are personnel of an extension organization (traditionally associated with agriculture) having roles in encouraging clients in the community to bring about their development and the development of society following a democratic approach. The role of extension is consistent with the leadership definition as a process, whereby an individual influences a group of individuals in the community to improve job performance and standards of living.

The importance of leadership role is also covered in the first chapter. The leadership role is analysed from three perspectives including that of extension agents, extension organization, and extension clients. The importance of leadership role to the extension agents is obvious in many ways. First, it is related to external and internal changes in response to the rapid development of global economy and complex social, economic, and environmental conditions. In such an environment, skills and knowledge quickly become obsolete and new skills and practices have to replace the old ones. Thus, the roles of extension agents are also transformed into a more complex and interconnected role, which needs different mindsets and ways of thinking. Their roles evolve from someone who merely delivers the contents of technological information in to a facilitator who facilitates progress, creates a vision, and builds capacities and teamwork, which are all undeniable roles in leadership. Second, the agricultural extension organizations are currently experiencing rapid changes and transition from functional hierarchy to being more customer-driven with a cost-effective approach that requires fast and flexible methods to meet the changing customer needs. It is the characteristic of learning organization that requires the concept of leadership to be expanded beyond the individual to the group functions in order to face any emergent challenge. Third, when agricultural extension agents are responsible for clients' empowerment, they must be empowered first before they can empower their clients. Empowerment of extension's clients would enable them to make their own decisions and solve their own problems, becoming less dependent on agricultural extension agents to manage their enterprises.

Chapter 2 covers the historical background of both motivation and achievement motivation theories. It is then followed by two theories related to achievement motivation that the authors believed are comprehensive and appropriate in explaining the concept and associated factors, namely, Bandura's social cognitive theory and Farmer's achievement motivation model. Bandura's theory tries to recognize the thinking processes that individuals go through in decision making for their behaviour by considering the reciprocal relationship of environmental factors on a person's cognition. Farmer's model further extended Bandura's theory by considering the pre-influence of background.

People's experiences in the workplace differ in many ways due to many personal characteristics, some of which are changeable, others that are less amenable to change. Chapter 3 delineates the background factors which are considered as primary dimensions of diversity and are less amenable to change, exerting significant impacts and shaping our basic self-image, sense of identity, and early learning experiences. Researchers often ignore the interdependence of these with other (secondary) dimensions of diversity, in which the latter are changeable personal characteristics that are acquired and may be modified or abandoned throughout life, which is discussed in Chapter 4. Although the secondary dimensions of diversity are more controllable, they will be affected by other people's decisions, which in this book we see as environmental factors, as covered in Chapter 5.

The reliability and applicability of the above-mentioned theoretical discussion have been tested in an original study conducted in Iran; its results are presented in the second part of the book. The second part of the book provides empirical evidence of factors influencing achievement motivation in the leadership role of extension agents. Past studies have indicated that there were variations in the results regarding extension agents' level of leadership role attainment, and little is known about the level of achievement motivation in leadership role among extension agents in Iran, a country location which had been chosen for the study. In Iran, extension system is the foundation of agricultural development. The agricultural sector is an important component of the Iranian economy, contributing 27% to GDP, 23.8% to employment opportunities, 82% to food supply, and 35% to non-oil exports. Extension agents in Iran are responsible for leading the clients, which, in 2009, consisted of 32 percent of Iran's total population of 66.5 million. This group of the population lived in rural areas and engaged in various types of farming. Chapter 6 covers the methodological perspectives of the study. Chapter 7 presents the results of factors influencing achievement motivation in the leadership role of extension agents. The interpretation of the study results, along with the conclusion, implications, and recommendations are presented in Chapter 8. This study is useful in designing a framework that improves our knowledge about achievement motivation in the leadership role of extension agents as the full mediation model of this study explained 78% of its variance. The study is a substantiation of Bandura's social cognitive theory by incorporating the pre-influence of background factors on personal and environmental factors that, according to Bandura, influence human behaviour. One of such a behaviour is achievement motivation. The practical contribution of this study comes from the fact that background factors' influences could be capitalized on in future learning of extension agents regarding how negotiation and support are made to reduce stereotypical assumptions on the leadership role. The influence of personal factors could assist extension organizations to develop appropriate training programs and could be used as a criterion in recruitment. Having employees who possess relevant personal attributes may not be sufficient to motivate them in leadership achievement if the environmental supports do not co-exist. Insights generated from this study would certainly be useful to improve performance of organizations, agents, and clients in the country's extension system.

The paragraphs that follow are set up to differentiate the originality of the present book in the area of achievement motivation studies and theories:

Literature on achievement motivation can be divided into three main groups of studies. The most frequent group is related to students and academic achievement motivation, which has some theoretical similarities but is different from the context of the present research. The second group is about predicting the career achievement motivation of students, such as the studies of Farmer (1985; 1997). The last group focuses on achievement motivation of higher positions and assigned role, such as achievement motivation of leadership role in managerial positions (London, 1997). A new insight, introduced in this book, is a specific focus on achievement motivation of leadership role which is an emergent role (discussed in Chapter 1) in extension context. Leadership role of extension agents is an emergent role which is achievable not because of the formal position in an organization, but because of the way clients respond to them. Emergent leadership appears over a period of time through communication and under the influence of different factors such as group member characteristics but the focus of this research is on the factors that motivate extension agents themselves to achieve and carry out the emergent leadership role. Achieving a leadership role would improve the individual and organizational performance of extension systems, and as a result would benefit clients as the target group of the extensions program.

Investigation of motivation and achievement motivation research and theories showed that most of the researchers and theorizers consider only one dimension of achievement motivation in defining and measuring achievement motivation. For example, status attainment theory just considers desire level of attainment (aspiration) to find how people achieve a status. The McClelland theory of achievement motivation mostly discusses mastery and desire to choose a challengeable but accessible task as achievement motivation. While according to Farmer, aspiration, mastery and salience all together cause achievement motivation (see Chapter 2). Thus, Farmer's model is given priority in defining achievement motivation due to the comprehensive point of view. Although Farmer worked on young subjects; the present book will adapt the multidimensional model of Farmer for the extension agents.

A particular aspect that had remained neglected when investigating achievement motivation, was considering the background factor that has pre-influence on personal and environmental factors, and most of the theories and researches are silent on this point (see Chapter 3). The background factors are considered as primary dimensions that shape our basic self-image, sense of identity, and early learning experiences. These dimensions fall into different categories whether it is regarded as not easily changeable or not changeable depending on society's belief and situations. Researchers often ignore the interdependence of them with secondary dimensions of diversity, which are personal factors.

Previous studies have integrated different variables within personal factors that motivate a person to achieve, for example, Leal (2006) investigated self-efficacy, work orientation, competitiveness, and personal unconcern. However, in extension context, the available research of motivation mostly investigated the demographic characteristics of extension personnel, such as age, gender, race, marital status, and education, as the personal factors (Ladebo, Olaoye, & Adamu, 2008). Therefore, a substantiation on the personal factors affecting achievement motivation of the agricultural extension is presented in Chapter 4.

The underlying assumption of environmental factors is derived from path-goal theory in leadership as the departure point of this factor. This theory emphasizes enhancement of employees' performance by focusing on their motivational sets and providing rewards in the work environment, though subjective types of support are neglected. The evaluation criteria for the environmental factor as incentives to achieve differ among the research contexts. For instance, the study of Okorley, Gray, and Reid (2009) highlighted the importance of fostering an open environment where extension staff feel comfortable to meet, interact, share information and ideas and are motivated to work through a participatory approach. Farmer argued for the importance of supportive environment, and Powell and Graves added the importance of supportive environment in terms of work and family friendly policies. Thus, in Chapter 5 a combination of Farmer and Powell and Graves' ideas are considered as criteria of determining environmental factor variables.

The book would be of interest to extension agents, extension service organizations, extension education students, extension scholars, and of course Iranian scholars. The discussion of background, personal, and environmental factors make the book interesting to any scholar of gender studies, psychology, and organizational behaviour as well.

PART I

ACHIEVEMENT MOTIVATION IN THE LEADERSHIP ROLE: A THEORETICAL CONSIDERATION

CHAPTER ONE

THE LEADERSHIP ROLE IN EXTENSION CONTEXT

This book is concerned with the achievement motivation of the leadership role among extension agents. The extension agent is someone who encourages clients in the community to bring about their development, as well as that of the society, following an appropriate approach in a democratic way, which is all about leading the clients using the leadership role. Thus, this chapter discusses the overview of the history of extension systems worldwide, including Malaysia as well as Iran, in which the latter is the context of the study. Further, this chapter discusses the changing demand being put on the public sector extension, globally and in particular in Iran, and suggests that these changes require extension agents to develop a leadership role of supporting and encouraging changes at farms and households. The importance of understanding achievement motivation in the leadership role of extension agents is in its relation to the increasing personal achievement which could further enhance organizational achievement by providing higher productivity, efficiency, and satisfaction. The achievement motivation in the leadership role of extension agents is also influential in extension's client achievement by empowering them in a leadership role through having a good role model as an extension agent who possesses the required leadership qualities.

History of Extension Systems Worldwide

The extension education originated in British universities around 1873, when Cambridge University started to do it, and the second was in Japan in 1893 who followed the concept and began to implement it (Ismail, 1990; Md.Sail, 2009). In general, agricultural extension worldwide has its roots as far back as 1955-1980, and aims to develop national agriculture extension systems, build new agriculture universities, and train indigenous extension workers (Thompson, 2002). The shift of the developmental program from agriculture to manufacturing and services in the middle of the 1980s to some extent renders the extension education work less important as a discipline that is traditionally associated with agriculture. In 1997 and 1998 Asian economic crises revived the interest in extension and the agricultural sector (Md.Sail, 2009). In the early years of the nineteenth century, extension services were in their formative stage in which they were relatively small in scale and limited in the scope of their work, only focussing on contact with farmers. They were organized by governments, and two main concerns at that time were fiscal stringency and relationship between research and extension. At the beginning of the twentieth century, extension organizations began to aim at broad national and farming system coverage of economic and social aspects (Ponniah, Puskur, Workneh, & Hoekstra, 2008). Thus, extension systems around the world are no longer a unified public sector aiming at only technical activities, but are a network of country institutions with multi-activity in multiple locations whose objectives and goals are derived from indigenous policy choices to support rural people (Rivera & Alex, 2004).

To revitalize agricultural extension for supporting people engaged in agricultural production and improving their livelihoods and well-being, international efforts during the past decade have provided a variety of institutional reform options. The term 'pluralistic' has been coined to capture this emerging diversity (Birner et al., 2009). The reforms were things such as decentralization, commercialization, privatization, non-governmental organizations (NGOs), farmer-to-farmer extension, participatory extension, group approaches, and gender specific extension (Van den Ban, 2009). While pluralism is still the subject of debate, the "pluralistic partnerships" appear as a viable and meaningful result of pluralism reforms, which are more likely to promote participatory approaches (Rivera, 2001). Participatory development in the world is essential due to its ability in empowering groups, communities and organizations by negotiating with institutions and bureaucracies (Rivera &Alex, 2004).

Extension systems in Asian countries (such as Malaysia and Iran) have a large number of similarities in which agricultural development and extension is under the responsibility of the ministry of agriculture. Transfer of technology is understood as the public funding for extension, the need for extension to engage with a wide range of issues beyond transfer of technology, and agricultural extension used to play an important role in promoting Green Revolution (Zhou, 2009). The extension systems in Malaysia and Iran will each be further clarified in the next sections.

History of Extension Systems in Malaysia

According to Md.Sail (2009), the extension work of Malaysia began in 1946 with the establishment of the Department of Agriculture (DOA) by the Colonial British government which focused on regulatory activities as providing agricultural input, and the advisory function in extension work was started from the middle of the 1970s. By shifting the focus of development in the middle of the 1980s from agriculture to manufacturing and services, the interests in the extension work were reduced. However, it was revived in 1997 and 1998 after the Asian economic crises and agriculture was introduced as the third engine of growth, and subsequently extension again became the focus of attention. Further, extension work was felt necessary in industrializing Malaysia, as well as for smallholders (Md.Sail, 2005; 2009).

The extension education was introduced as an academic subject in 1967 in the College of Agriculture Malaya. The Department of Extension and Home Economics was established in 1971 in the same year with the establishment of University of Agriculture Malaysia (UPM). To respond to the need for linking research, extension and clients in technology dissemination and adoption, UPM formed the Center for Extension and Continuing Education with two academic departments, one of which was the Department of Extension Education. In 1996 the department was relocated to the current Faculty of Educational Studies. The department of Extension and Education, in addition to serving national and international students in Master and PhD degrees, provides extension services to farmers, fishermen and other rural populations, as well as on the job training services to extension staff (Md.Sail, 2009).

The agricultural extension services in Malaysia are placed under the jurisdiction of separate organizations (Md.Sail, 2002, 2005) such as the Department of Agriculture (DOA), the Rubber Industry Smallholders Development Authority (RISDA), the Federal Land Development Authority (FELDA), the Federal Land Consolidation and Rehabilitation Authority (FELCRA), some regional development agencies, and the state land development agencies. The Malaysian extension system is based on the commodity approach, providing specific services to rubber, oil palm, or livestock sectors. Further, there are divisions according to the functional activities such as research, marketing, farmers, and according to specific areas and commodities such as Malaysian Agriculture Research and Development Institute (MARDI), Malaysian Palm Oil Board (MPOB), and Muda Agriculture Development Authority (MADA). The specialization has its advantage in providing a more efficient system, while the disadvantages are in providing services to the same group of clients who have livestock and other cash crops, duplication, and wastage of resources (Md.Sail, 2005).

The concept of extension in Malaysia includes two traditions of transfer of technology (ToT) and human resource development (HRD). The ToT process, which is the primary focus of extension in Malaysia, starts from researchers who generate and develop appropriate technology for a specific location as mature technology, then passing it down to extension agents to repackage it to a simpler form, and disseminate to clients using teaching and learning methods. The HRD tradition is related to the adult education model and focuses on the analyses of client needs, problems, and situations (Md.Sail, 2002).

History of Extension Systems in Iran

The non-formal extension work in Iran dates back to two centenaries ago, and was started by writing a book about what today is entitled as research, extension, and education. Further, there had been some newsletters attempting to disseminate new technologies to increase farmers' productivity (MalekMohammadi, 2000). But the formal extension services actually began more than 50 years ago (Karbasioun & Mulder, 2004).

The Ministry of Agriculture in Iran has been established since 1941, and the staff of the Provincial Agricultural Headquarters handled the rural development affairs. After World War II, the predominant conditions drew the attention of the world in general and Iran in particular, to the agricultural extension activities. Thus, the Agricultural Extension Department was established in 1948 as a department under the Ministry of Agriculture. In 1953, agricultural extension was located within the framework of an independent instructional organization with the support of the Ministry of Agriculture. Although the Agricultural Extension Department in the first decade of its activities merely worked on educational matters, in 1962 it concentrated on the implementation of the Land Reforms Program and in 1973 the main focus was the implementation of projects to increase agricultural production. During these three decades the main activities of the extension department were provision of loans and credit, preparation and arrangement of works, handling and distribution of seeds (Ghareyazie, Aghajani, & Ranjbar, 2003). One could argue that from the initial launching of extension to the end of the nineties, extension workers played the role of experts in the transfer of technology, and the role of extension agents as the ones who develop human resources, was neglected. Further, Fami (2006) felt the lack of an actual linkage between farmers and scientists to develop human resources (farmers) at that time.

Afterwards for developing meaningful relations between research, extension and farmers (both in quality and quantity), the Agricultural Research, Education and Extension Organization (AREEO) emerged in 1993 as one of the dependent organizations of Jihad-e-Agriculture Ministry. Then, the main functions of extension are defined as: First, functioning on the basis of public extension attending the target groups (rural women, rural youth, etc.). Second, focusing on agricultural projects aimed at strengthening the provision of credits. Third, focusing on technical/professional extension and transfer of research findings (Ghareyazie et al., 2003). At this time, the importance of extension agents' leadership role with consideration of similarity between leader and client emerged as a main competency to perform the extension function, especially for the various target groups of rural men and women.

Overall, Iran's agricultural extension services in the current decade have been going forward with a noticeable rate in all enclosure roles including transfer of technology and human resource development; however, the human resource development role of extension agents have not been considered satisfactorily (Karbasioun & Mulder, 2004). Furthermore, farmers accepted the agricultural extension service as the most-used source of information, and extension services as a non-formal education provides advisory services to the clients, followed by a combined approach with a focus on the ministry-based extension system. Other approaches are commodity-based for strategic crops such as wheat, project-based such as watershed management extension, client-based such as women or youth income-generating projects, university-based, training and visit (T&V), participatory, and privatized extension as a more recent approach (Karbasioun, Biemans, & Mulder, 2007).

Role of Extension Agents in Iran

Iran's extension agents are responsible for providing the knowledge and information that will enable a farmer to understand and make a decision about a particular innovation. The knowledge provided for a farmer is mostly technical knowledge based on the structured programs of Ministry and government policies for rural development. The main concern of mentioned responsibility is the transfer of technology that creates the role of technical leader for extension agents to the clients. MalekMohammadi (2000) also discussed the role of Iran's extension agents as technical leaders. He believed that the achievement of a technical leader role of extension agents takes time and needs an increasing effort to create a trustful situation in which rural people accept the extension agents as technical leaders. Rural people accept someone as their technical leader only when they see his or her technical excellence, along with seeing their local leader following that person. Thus, extension agents, in order to confirm their role as technical leader in addition to possessing the appropriate amount of technical knowledge and skills, need local leaders' support. Giving deep respect to the local leaders, which vary in rural areas such as a religious leader as the predominant one, would increase the respect of extension agents as technical leaders. It is consistent with the reviewed literature on the principle of early extension work (Franco, 1960) with emphasis on using a local leader.

Due to current emphasis of scholars and extension organizations on a participatory approach of extension, the extension agents should act as facilitators to help and encourage farmers to develop their own initiatives and to begin to tackle their own problems, and consequently farmers' personal development (Shirzad, 2006). It is consistent with the recent literature on the principle of extension on leadership development (Nisha, 2006) and definition of leadership (Northouse, 2007) which is all about clients' empowerment. Thus, recently there are more emphases on human resource development rather than the transfer of technology that brings the significance of the role of extension agents as holistic leaders, not only technical leaders. In agreement with Morse, Brown, and Warning (2006), the role of extension agents was to improve the practices on the farm, in the home, and in the community. According to a study conducted in Iran by Karbasioun, Biemans, and Mulder (2007), the farmers claimed that agricultural extension services have supported them to some extent on animal husbandry and veterinary matters, agricultural input and enhancement of the fertility and size of the farms, while for the future they expressed the need for greater support regarding socialization competencies, emotional aspects, and making an agricultural career more satisfactory. While the global environments have evolved, the roles of extension agents have also become more complex and interconnected, in which the "programming" approach, which is one of the predominant approaches of extension system in Iran, is not sufficient. For the extension agents to work effectively in the future, they need a different way of thinking in order to act as a facilitator to make a difference.

Iran's extension agents could hold different organizational positions, from technician to a top manager, in order for them to play the different roles. The first organizational position of extension agents is being a technician who has a technical educational qualification. The second level is being expert, where the holders have a Bachelor's degree. The extension agents with technician and expert positions are a life line of extension services and they are directly engaged with training of clients. The third level is the responsible expert whose duties are similar to the subject matter specialist. Further, a person in charge of such a position mostly possesses a Master's degree in extension education. He/she is responsible for a specific subject or project and gives related advice to the clients. The fourth level is the field level managers, called the first managerial level. The person in charge of this position, in addition to giving advice to the clients, is the supervisor of an office in extension organization. The fifth positional level of extension organization is the middle (second) managerial level in which the person is responsible for supervising several offices in extension organization. The highest (top) managerial level is the position of a dean in the extension organization of provinces. In sum, employees who carry the name of extension agents in any organizational level or position are responsible for establishing communication and contact with clients. It is where the emergent leadership role of extension agent to the client appears.

Emergent Leadership Role of Extension Agents

An extension agent is someone whose job is to approach clients and in which they perform tasks beyond merely being a technical innovation motivator to human resource development leaders, helping in mobilization of resources in the community (Khalil, Ismail, Suandi, & Silong, 2009). Leadership role of extension agents is an emergent role that is achievable, not because of formal position in an organization, but because of the way clients respond to them. Emergent leadership appears over a period of time through communication and under the influence of different factors, such as group members and leader characteristics (Northouse, 2007). Furthermore, the differences of leadership role from managerial position rely on a relationship and a function. Management is a function of planning, budgeting, evaluating, and facilitating, while leadership is a relationship for selecting talents, motivating, coaching, and building trust (Maccoby, 2000).

Philosophically, the role of extension workers is to encourage clients in a community to bring about their development, and that of the society following an appropriate approach in a democratic way (Chauhan, 2007). The overall role of extension is consistent with the leadership definition as a process whereby an individual influences a group of individuals in the community to achieve a common goal in a democratic way (Northouse,

2007). Thus, it could be concluded that extension philosophy is all about influencing or leading clients to achieve clients' wellbeing. When the social, economic, and political situations have evolved, the roles of extension workers also change into a more complex and interconnected role, which needs different mindsets and different ways of thinking. The role of extension workers evolved from someone who merely delivers the contents of technological information to a facilitator that facilitates progress, creates a vision (Morse et al., 2006), and builds capacities and teamwork (Vijayaragavan et al., 2005), which are all criteria of leadership role.

The extension work is based upon extension principles as well. The knowledge of these principles is necessary for an extension worker. Analyses of scholarly writings about the extension principles show that the "use of local leaders" as one of the most important principles of extension (Franco, 1960) was explained in past literature, while in recent literature, the term has been replaced by "leadership development" (Nisha, 2006). Leadership development among rural people relates to a process of strengthening groups by enhancing the capacity of members to work together effectively (Vijayaragavan et al., 2005). Although the "use of local leader" has always been an important principle (MalekMohammadi, 2000), the differences of past and present scholarly writing mainly rely on the development of leadership abilities among clients instead of just using the available leaders. However, the emphasis on leadership development of extension agents who act as the role model to the rural people is always missing. Although there is increasing attention to keep extension professionals updated and motivated, Chauhan (2007) has expanded the principles of extension by considering the "principle of trained specialists" and "principle of workers encouragement". But as yet there is a lack of studies on factors influencing achievement motivation of extension agents' leadership role that could help to update the fact that extension workers act as role models to the clients, which is the focus of this book.

The following sections discuss the importance of leadership role in the current situation of extension systems from three perspectives, including; extension agents, extension organization, and extension clients.

Importance of Leadership Role to the Extension Agents

The importance of leadership role to the extension agents or extension organization employees is related to external and internal changes in response to the rapid development of a global economy and complex social, economic, and environmental conditions (Ladewig & Rohs, 2000).

In such an environment, skills and knowledge quickly become obsolete and new skills and practices have to replace the old ones (Vijayaragavan, Singh, & Wason, 2005). Successful performance of new roles calls for efficient extension professionals with skills of modern management practices, such as leadership (Vijayaragavan et al., 2005). The leadership role nowadays is used as part of extension agents' performance appraisal (Fabusoro, Awotunde, Sodiya, & Alarima, 2008) and one of the required competencies of extension agents (Vijayaragavan & Ghosh, 2006). The leadership role of extension agents will become even more important as resources and staff become more limited and tasks become varied in a new era of change (Ladewig & Rohs, 2000). Further, the importance of motivational factors of leadership achievement is supported through extension principles (Chauhan, 2007).

The context of agricultural extension has changed in recent years. Today's extension is seen, not only in terms of greater opportunities through new technologies and globalization, but also challenges due to population growth and environmental issues. Some of the opportunities are due to recent developments in sustainable technology, biotechnology, modern communication, information technologies, economic liberalization, and globalization. The challenges faced are achieving increased productivity along with sustainability in the midst of poverty, population explosion, unemployment, poor institutional reform, environmental degradation, and ill effects of globalization (Vijayaragavan et al., 2005).

Karami, Chizari and Farhadiyan (2003) discussed the recent changes influencing Iran extension organization's capability to carry out its mission and goals. They asserted that rapid technological changes, financial limitation, and changeable needs of extension clients may appear as big threatening factors if the organization is not ready to answer these changes. Thus, extension organizations need to become more customer-driven with a cost effective approach, which requires fast and flexible methods in order to meet changing customer needs and continue to satisfy customer expectations (Ladewig & Rohs, 2000).

Consequently, agricultural extension services worldwide are facing one of the most challenging times and are confronted with the challenge of determining and implementing the necessary structural and organizational changes to ensure sustainability in the future (Sadighi, 2003) that will certainly affect the employees' roles. It is where the skills of extension agents become outdated and new skills and competencies, such as leadership, to address these changes are needed (Vijayaragavan et al., 2005). The role of the extension agents in the new environment of organizational transition are team facilitator, counsellor, coach, and encourager, rather than just an expert, overseer, enforcer, and judge (Patterson, 1998). These roles are consistent with the leadership role requirements, which are: managing job, building the team, relating to others, and thinking clearly (Ladewig & Rohs, 2000). Patterson (1998) suggested leadership as a new way of administrating in the workplace that will lead to new levels of productivity and satisfaction for extension organizations, extension workers, and extension clients.

Although the term "management" is noticeably absent from most extension literature, the overriding activity of extension administrators (or even field level extension workers) is, primarily, to manage or to lead (Patterson, 1997). However, the leadership role of extension agents is very much emphasized in the literature. For example, FAO (2006) listed a broad variety of terms used in extension to explain the role of extension agents, of which all of them are about leading the clients. The roles are as a teacher, educator, communicator, facilitator, broker, consultant, intermediary, listener, organizer, motivator, leader, administrator, enabler, activist, provider, arbitrator, advocate, catalyst, friend, research linker, and stimulator.

Another issue that highlighted the importance of leadership role from employees' perspective is extension professionals' role conflict. Sasidhar, Rao, and Sreeramulu (2008) discussed the increased and varied activities of new environment, demanding extra time and energy, causing role-conflict and affecting job performance of extension professionals. If the factors of role conflict are not properly coped with, extension professionals' performance can be affected in a dysfunctional style. Furthermore, according to Ladewig and Rohs (2000) the leadership competencies will become even more important as tasks become varied, and resources and staff become more limited. To be successful in this new era of changes, extension workers must be able to manage tasks and to lead people.

The changing situation of extension system, which causes changes in the role of extension agents toward also tackling leadership role, was supported by the idea of Leeuwis (2000) about re-conceptualizing participatory rural development towards effective negotiation. To be truly effective in sustainable rural development, the participatory approach is a leading principle in which inability of facilitators to handle the conflicts causes its poor productivity. Negotiation is an alternative for organizing participatory development that requires different tasks, skills and roles for facilitators of participatory process such as extension agents (Leeuwis, 2000). Further, present thought on the extension system is a pluralistic demand-driven approach that has created new roles for both parties; the demand side as farmers and the supply side as extension agents (Klerkx &

Leeuwis, 2009). The concept of leadership in extension context involves all leaders, not only executive leaders, but also front-line workers and professional staff, who disseminate ideas throughout and outside the organization. All have essential roles in a new kind of teamwork to bring about client development (Rivera & Alex, 2004).

Fabusoro et al. (2008) suggested leadership attainment as one of the four required functional categories for performance appraisal of extension workers. The others are job assessment ability, character traits, and work habit. Vijayaragavan and Ghosh (2006) asserted the leadership ability as one of the critical qualities needed for the field level and supervisory extension staff. Movahedi (2008) further revealed the leadership ability for Iran agricultural extension students as one of the personal attributes that goes with motivation, confidence, and dedication for them to be successful in their careers.

Vijayaragavan, Singh, and Wason (2005), in their study for designing and validating training modules for improving management skills of extension professionals through task analysis and self-assessment, revealed the following areas of training needs: planning, leadership, communication, teambuilding, creativity, decision making and problem solving, personal effectiveness, motivation, internet for extension, stress management, time management evaluation of programs, and performance appraisal. Sulaiman and Ban's (2003) results also showed that extension staff need an increasing level of social skills related to group formation, leadership development, conflict resolution, and inter-group negotiation for them to work effectively in mobilizing and organizing farmer groups.

Thus, for the groups of extension workers that want to be more effective in responding to the rapid changes in today's environment, an appropriate mix of skills, including leadership, is needed. Clearly achieving extension agents' leadership role is something that goes further than employees' perspectives and also needs to be considered from an organizational perspective.

Importance of Leadership Role to the Extension Organizations

In this section the importance of the leadership role is investigated in terms of organizational transition. Agricultural extension organizations are currently experiencing rapid changes and transition (Khalil, 2008). The changes involve the use of new sustainable technologies in the farming system, such as high-yielding varieties of paddy, biotechnology, modern communication parallel with economic liberalization, population explosion, and solutions toward poverty and unemployment (Vijayaragavan et al., 2005). Thus, extension organization structure needs to be transformed from functional hierarchy to being more customer-driven with a cost effective approach, which requires fast and flexible methods to meet the changing customer needs. It is the criteria of learning organization that requires the concept of leadership to be expanded beyond the individual to group functions in order to face any emergent challenge (Allen & Morton, 2006). Extension organizations experience a new environment that changes motivational factors. Thus, investigation of extension agents' achievement motivation in leadership role becomes an organizational matter (Sadighi, 2006). To effectively make the transition, for instance, there is an organizational change from a functional hierarchy to a learning organization. A learning organization emphasizes the supplementary role of leadership practice for bureaucracy (Ladewig & Rohs, 2000). A learning organization requires the concept of leadership to be expanded beyond the individual to group functions. This means that both manager and employee need to practice leadership role in learning organization (Allen & Morton, 2006).

Organizational or any kind of change or transition begins with a problem or a need that somebody really cares about; a recognition that something is wrong, that something requires change (Havelock & Hamilton, 2003). The external and internal changes that affect extension organization show the importance of organizational transitions that extension scholars care about (Aghasizadeh, 2003; Ladewig & Rohs, 2000; Patterson, 1998). Since social systems do not automatically welcome such sudden desire to change, the recognition of transition and required changes, such as employees' roles, must be escorted by feelings of necessity. Thus, to initiate a change, the investigation of the motivational set, such as achievement motivation of leadership role among employees as a predominant role in new context, is vital for extension organizations (Havelock & Hamilton, 2003).

The challenges facing extension today stem from the basic problems of the extension systems, which are bound to its diverse functions, as well as the bureaucratic, political and social operating environments within which extension systems function (Fabusoro et al., 2008). This is consistent with Sadighi's (2003) contention about highly bureaucratic and ineffective organizational structures that are typical shortcomings of conventional extension organization common in many developing countries. Thus, extension organizations need to move from diverse and supportive functions in a bureaucratic environment toward being a learning organization. The leadership role of extension workers will become even more important due to changes in the availability of resources in the farming system (Ladewig & Rohs, 2000).

Rural land managers need access to sound advice and information to respond to pressures from environmental regulations, declining farm incomes, changing patterns in international trade, and new institutional arrangements within the domestic food chain (Garforth, Angellb, Archerc, & Green, 2003). Thus, extension systems should go beyond the straightforward provision of advice and information, to processes of confidence building, facilitation of learning, and motivation. Further, the authors emphasized the term facilitation, which covers a wide range of support, including assistance to an individual farmer in completing an application for funding, to the building up of leadership capacities in communities (Garforth, Angell, Archer, & Green, 2003).

Pretty and Vodouhê (2005) discussed the most important challenge of agricultural organizations to become learning organizations. The transition to learning organization further implies new roles for extension professionals, leading to a whole new professionalism with new organizational structure. Havelock and Hamilton (2003) discussed the three-step model of transition in social systems that includes system openness, moving, and refreezing. One could argue that the extension organizations in Iran are in the middle of system openness for transition (Sadighi, 2006), which shows the importance of practicing leadership role in undergoing the transition. Cristovao et al. (2009) asserted the importance of helping organizations to learn, that is, to become 'learning organizations' through leadership practice.

Appropriate response to the changing nature of work and authority relationships is the expected goal of organizational transition that emphasizes extension agents' leadership role (Ladewig & Rohs, 2000). Although the target of extension organizations' transition is entitled differently, such as entrepreneurial organization (Rivera & Davis, 2008), self-organizing capacity (Allen & Morton, 2006), and learning organization (Ladewig & Rohs, 2000), all of them highlight the importance of extension agents' leadership role in new organizational structure. Thus, to perform a leadership role, extension organizations have to reorganize their structure by embracing wider expertise, decentralizing management, and nurturing a culture of learning organization (Sulaiman & Ban, 2003). A learning organization is concerned with a mind shift from seeing parts to seeing wholes, from seeing people as helpless reactors to seeing them as active participants in shaping their reality, from reacting to the present to creating the future (Gruidl & Hustedde, 2003). Leadership development stimulates individuals to think critically and to recognize innovative solutions to achieve present and future goals. While organizations are inspired to become creative and innovative, decentralized forms of leadership that engage teams of people working on common projects become more critical. Organizations that develop and support shared leadership among groups of members are more expected to let individuals at all levels be involved in decision-making and actions that benefit the whole, and they are learning organizations (Allen & Morton, 2006).

Morris (2003) believed that the information revolution and a new focus on communities provided an opportunity for a richer, more participative democracy than ever before. Thus, extension agents need to consider the option of truthfully engaging with their community and producing the opportunity for a richer, more participative extension than ever before. One could argue perhaps that a focus on leadership role of extension agents is one of the opportunities for extension agents, due to the need for a more participatory way of managing the task and truthfully engaging with the clients.

Importance of Leadership Role to the Extension Clients

The third perspective on the importance of extension agents' leadership role is related to the extension clients. For an extension worker to be successful in this new era of changes, they must be able to manage tasks and to lead clients. Nowadays, public sector extension has to provide information and advice to farmers, not only on technologies, but also on solutions to specific field problems, markets, prices, quality, and implications of policy changes. Public sector extension also has to organize farmers to take collective action to adopt new technologies, to solve problems, and to increase income from agriculture (Sulaiman & Ban, 2003). Gruidl and Hustedde (2003) discussed the idea that extension organizations strengthen groups by enhancing the capacity of members to work together effectively through leadership practice. In the coming years, agricultural extension will have to go further than the role of mere transfer of technology, like getting involved in new roles related to capacity building, team building, and institutional restructuring (Vijayaragavan et al., 2005). Therefore, the role of extension agents to the clients is to empower them in farm productivity and management, as well as their wellbeing and that of the society.

Empowerment is a relatively new concept in an extension context. Empowerment refers to capacity building process, and sharing knowledge and skill through a teaching and learning process. Empowerment comprises structural and psychological domains. Structural empowerment refers to practices and policies, and psychological empowerment refers to the sense of being empowered. Thus, when agricultural extension agents are responsible for clients' empowerment, they must be empowered first before they can empower their clients. If agricultural extension agents are empowered, then they will in turn empower their clients. If clients are empowered, then they would be able to make their own decisions and solve their own problems, becoming less dependent on agricultural extension agents to manage their enterprises (Md.Sail, & Asnarulkhadi, 2010). The sense of belongingness and ownership are significant benefits of empowerment (Md.Sail, & Asnarulkhadi, 2010), and decentralization of leadership role (Duncan, Navarro, & Mykerezi, 2008) from extension agents to the clients, acquired through teaching and learning processes, which are important concepts of the human resource development function of extension agents (Md.Sail, & Asnarulkhadi, 2010). Thus, the human resource development function of extension agents includes a leadership role, which is a platform for reforms and client empowerment that is highly emphasized in the new global situation (Rivera & Alex, 2004).

The importance of leadership development in rural areas is also due to the presence of a skilled leader as one of the criteria for agricultural cooperative success (Androulidakis, 2001). Regarding economic and agricultural development, particularly in developing countries, Goff, Lindner, and Dooley (2007) introduced cooperatives as the best-suited institutions, in a study assessing the perception of three coffee cooperatives' boards of directors, management, and membership regarding their cooperative's organizational and leadership development. They believed that achievements of agricultural cooperatives depend on democratic processes and bottom-up decision making. Bottom-up decision making is about improving the leadership competencies of extension clients which is considered as a criterion for long-term sustainability. One could argue that a cooperative, as a participatory approach concerned with establishing farmers as a group, is one of the main vehicles in developing clients so as to develop the agricultural systems. Thus, extension agents, by leadership function in a participatory climate, offer the right opportunities of learning to the clients and help their personal development instead of simply supporting them. This is consistent with the ancient Chinese proverb, "Give a man fish he will have food for the day; Teach him how to fish he will have food for the rest of his life". The key point related to this explanation is that while leadership traits of extension clients are criteria for sustainability, the extension agents, as the role models of these traits for the clients, have to possess a leadership role earlier than the clients.

Cooperative extension is poised to make community action leadership development its hallmark for the 21st century, and its unique and timely contribution to the burgeoning field. But extension staff gave less emphasis to dealing with change, diversity, and conflict, transformational or visionary leadership (Sandmann & Vandenberg, 1995). Thus, extension workers who need to work in real sympathy with their clients, are often less likely to take a leadership role that truly transforms their organization to compete in the new era of changes (Vijayaragavan et al., 2005). Because of this concern, many extension organizations are struggling to develop leadership competency of extension agents and an organizational climate that can enable the group of extension agents to respond to change (Allen & Morton, 2006; Ladewig & Rohs, 2000).

Conclusion

The agricultural extension climate in many developing and developed countries is influenced by the quality of the extension staff. The field level extension agents are the lifeline of the agricultural extension system. Although they appear to be the last in the chain of command, they are perhaps the most important element in the system of extension; their motivations to perform their job are vital in agricultural development. One could argue that extension agents' achievement motivation in leadership roles is also vital, as was discussed earlier from the three perspectives.

In sum, the agricultural extension organizations are currently experiencing transition and rapid changes. In such an environment, personnel motivation changes as time and conditions change; furthermore, skills and knowledge quickly become expired and new skills have to replace the old. Successful performance of new skills in an extension context calls for efficient extension professionals with leadership competency. To effectively make the transition from a functional hierarchy to learning organization, leadership practice would supplement bureaucracy. Learning organization also requires the concept of leadership to be expanded beyond the individual and moves toward group functions. The leadership role of extension agents will become even more important as resources and staffs become more limited and tasks are varied in the new era of changes. Extension agents' leadership roles would also be able to develop clients' capacity: a procedure to empower the clients from being dependent on extension agents to being a decision maker. Therefore, leadership role attainment is considered one of the main competencies required by extension agents and instrumental for extension agents' performance appraisal.

Extension scholars are concerned about the gap regarding the taking of a leadership role by extension agents. Although many extension organizations are struggling to develop leadership among extension agents through developmental programs, the results are not yet acceptable. Agricultural extension in the coming years has to go beyond the role of mere transfer of technology and get involved in leadership roles related to capacity building, team building, and institutional restructuring. Studies have indicated that extension professionals lack skills in the diverse areas of management, such as leadership, which is of importance for their efficiency and productivity in the new era of change. Studies have declared that extension professionals choose not to take on volunteer leadership roles, and have found leadership abilities to be the component in which respondents need more on-going education and training. Thus, it is suggested that, in order to work effectively, extension staff also need an increasing level of social skills, such as leadership. Furthermore, there are variations in the study results regarding extension agents' level of leadership role attainment. Most importantly, little is known about the level of motivation for attainment of leadership role among extension agents in Iran. Thus, this book, after presenting a theoretical discussion of the concept of achievement motivation, intends to go back to the basic steps so as to investigate what factors influence the achievement motivation in leadership role among extension agents.

References

Aghasizadeh, F. (2003). *Reconstructing of Iran extension organization structure according to requirement of new alteration and its need (doctoral dissertation).* Ph.D., Azad University Unit of Tehran Research and Science Tehran, Iran.

Allen, B. L., & Morton, L. W. (2006). Generating self-organizing capacity: leadership practices and training needs in non-profits. *Journal of Extension, 44*(6).

Androulidakis, S. (2001). Leadership ability dimensions in relation to rural women's personal characteristics: A methodological model for further consideration. *Journal of International Agricultural and Extension Education 8*(1), 39-46.

Birner, R., Davis, K., Pender, J., Nkonya, E., Anandajayasekeram, P., Ekboir, J. A., Cohen, M. (2009). From best practice to best fit: A framework for designing and analyzing pluralistic agricultural advisory services worldwide. *Journal of Agricultural Education and Extension, 15*(4), 341-355.

Boyd, B. L. (2004). Extension agents as administrators of volunteers: competencies needed for the future. *Journal of Extension, 42*(2).

Chauhan, J. (2007). Introduction to Agricultural Extension Education *Agricultural Extension Education*. India: R. B. S. College, Bichpuri.

Cristovao, A., Ferrao, P., Madeira, R., Tiberio, M. L., ′nho, M. J. R., & Teixeira, M. S. (2009). Circles and communities, sharing practices and learning: looking at new extension education approaches. *Journal of Agricultural Education and Extension, 15*(2), 191-203. doi: 10.1080/ 13892240902909122

Duncan, D. W., Navarro, M., & Mykerezi, P. K. (2008). *Learning about sustainability: the contribution of the global seminar educational model.* Paper presented at the 24nd Annual Conference of AIAEE "Global Entrepreneurship in International Agricultural and Extension Education", March 9 to15, Costa Rica.

Fabusoro, E., Awotunde, J. A., Sodiya, C. I., & Alarima, C. I. (2008). Status of job motivation and job performance of field level extension agents in Ogun State: Implications for agricultural development. *Journal of Agricultural Education and Extension, 14*(2), 139-152.

Fami, H. S. (2006). *Islamic Republic of Iran* Paper presented at the Enhancement of Extension Systems in Agriculture, Published by the Asian Productivity Organization, Pakistan, 116-125.

FAO. (2006). Guide to extension training Retrieved october 10, 2009, from <http://www.fao.org/docrep/t0060e/T0060E08.htm>

Franco, J. D. (1960). *Extension philosophy*. Costa Rica: Inter-American Institute of Agricultural sciences, Turrilba.

Garforth, C., Angell, B., Archer, J., & Green, K. (2003). Improving farmers' access to advice on the land management: lessons from case studies in developed countries. *Agricultural Research & Extension Network, Network Paper No. 125.*

—. (2003). Fragmentation or creative diversity? Options in the provision of land management advisory services. *Land Use Policy, 20*(4), 323-333.

Ghareyazie, B., Aghajani, M., & Ranjbar, G. A. (2003). *Integration of agricultural research and extension in Islamic Republic of Iran.* Paper presented at the Integration of Agricultural Research and Extension, March 18 to 22, Published by the Asian Productivity Organization, Philippines.

Goff, S., Lindner, J. R., & Dooley, K. E. (2007). *Organizational and leadership development: A case study of three coffee cooperatives in*

Rwanda. Paper presented at the Internationalizing with Cultural Leadership, May 20 to 24, Polson, Montana, Polson, Montana.

Gruidl, J., & Hustedde, R. (2003). Evaluation of capacity-building programs: A learning organization approach. *Journal of Extension, 41*(5).

Havelock, R., & Hamilton, J. (2003). *Guiding change in special education: how to help schools with new ideas and practices*: Corwin Press.

Ismail, M. (1990). *Pengembangan: Implikasi ke atas pembangunan masyarakat (Extension: Implication to community development).* Kuala Lumpur: Dewan Bahasa dan Pustaka.

Karami, R., Chizari, M., & Farhadiyan, H. (2003, 2003). *Iran's extension organization, challenges and opportunities.* Paper presented at the First Conference of Iran's Agricultural Extension, Education and Natural Resource, Tehran, Iran.

Karbasioun, M., Biemans, H., & Mulder, M. (2007). Supporting role of the agricultural extension services and implications for agricultural extension instructors as perceived by farmers in Esfahan, Iran. *Journal of International Agricultural and Extension Education, 14*(1), 31-44.

Karbasioun, M., & Mulder, M. (2004). *HRM and HRD in agricultural extension organizations in Iran: A Literature Review.* Paper presented at the 20th Annual Conference of AIAEE, Dublin, Ireland, pp. 13-24.

Khalil, A. H. O. (2008). *Relationships between extension competencies, organizational commitment and job satisfaction with performance of agricultural extension workers in Yemen.* Doctoral dissertation, Faculty of Educational Studies, Universiti Putra Malaysia, Serdang, Selangor.

Khalil, A. H. O., Ismail, M., Suandi, T., & Silong, A. D. (2009). Human resource development competencies as predictors of agricultural extension agents' performance in Yemen. *Human resource development international, 12*(4), 429-447.

—. (2008). Extension worker as a leader to farmers: influence of extension leadership competencies and organizational commitment on extension workers' performance in Yemen. *The Journal of International Social Research, 1*(4), 368-387.

Klerkx, L., & Leeuwis, C. (2009). Operationalizing demand-driven agricultural research: Institutional influences in a public and private system of research planning in the Netherlands. *Journal of Agricultural Education and Extension, 15*(2), 161-175.

Ladewig, H., & Rohs, F. R. (2000). Southern extension leadership development: Leadership development for a learning organization. *Journal of Extension, 38*(3).

Lamm, A., Nistle, D., & Stedman, N. (2010). *Identifying the influencing factors affecting professional volunteer leadership in extension* Research Paper Retrieved from www.leadershipeducators.org/.../Lamm-Nistler-Sredman%20FINAL.PDF

Leeuwis, C. (2000). Reconceptualizing participation for sustainable rural development: toward a negotiation approach. *Development and Change, 31*, 931-959.

Maccoby, M. (2000). Understanding the Difference Between Management and Leadership. *Research Technology Management, 43*(1), 57-59.

MalekMohammadi, I. (2000). *Agricultural extension education and human resources*. Tehran, Iran: Tehran University Publication.

Morris, D. (2003). *Tasmania together: new democracy at work.* Paper presented at the Extending extension: beyond traditional boundaries, methods and ways of thinking, November 26 to 28, Hobart, Tasmania, Australia.

Morse, R. S., Brown, P. W., & Warning, J. E. (2006). Catalytic leadership: reconsidering the nature of extension's leadership role. *Journal of Extension, 44*(2).

Movahedi, R. (2008). Agricultural extension education in Iran. Retrieved 21 Sep 2009, from www.movahedr.blogfa.com

Nisha, M. (2006). *Understanding extension education*. New Delhi, India: Kalpaz Publications.

Northouse, P. G. (2007). *Leadership theory and practice* (fourth ed.). Thousand Oaks, London, New Delhi: SAGE Publication.

Patterson, T. F. (1997). Fundamentally flawed: extension administrative practice part 1. *Journal of Extension, 35*(6).

—. (1998). A new paradigm for extension administration *Journal of Extension, 36*(1).

Ponniah, A., Puskur, R., Workneh, S., & Hoekstra, D. (2008). *Concepts and practices in agricultural extension in developing countries: A source book*: IFPRI (International Food Policy Research Institute), Washington, DC, USA, and ILRI (International Livestock Research Institute), Nairobi, Kenya.

Pretty, J. N., & Vodouhê, S. D. (2005). Using rapid or participatory rural appraisal. In B. Swanson & R. P. Bentz (Eds.), *Improving agricultural extension. A reference manual*: Daya Publishing House, FAO.

Md.Sail, R. (2002). *Adoption of research results: effectiveness of transfer of technology to rubber smallholder.* Paper presented at the IRRDB Liaison Officers, 8 January, De Palma Hotel, Kuala Lumpur.

—. (2005). *Towards an effective extension delivery system for rubber smallholders in Malaysia*. Paper presented at the International Rubber

Research and Development Board TOT Workshop on Smallholders, April 4 to 8, Kuala Lumpur, Malaysia.

—. (2009). *Revitalizing extension education for industrializing Malaysia: trends, priorities and emerging issues*. Paper presented at the Revitalizing Extension and Continuing Education: Linking University, Ministry, Industry, and Community, 26 October, IOI Marriot Hotel, Putrajaya.

Md.Sail, R., & Asnarulkhadi, A. S. (2010). Community development through community capacity building: A social science perspective. *Journal of American Science, 6*(2), 68-76.

Rivera, W., & Alex, G. (2004). Extension System Reform and the Challenges Ahead. *The Journal of Agricultural Education and Extension, 10*(1), 23-36.

Rivera, W. M. (2001). Agricultural and Rural Extension Worldwide: Options for Institutional Reform in the Developing Countries. Rome, Italy: FAO.

Rivera, W. M., & Davis, K. E. (2008). *Designing agricultural education systems that promote and prepare entrepreneur.* Paper presented at the 24nd Annual Conference of AIAEE "Global Entrepreneurship in International Agricultural and Extension Education", Costa Rica.

Sadighi, H. (2003). *Extension re-organizational engineering commensurate with progresses in technology.* Paper presented at the 19 Annual Conference of AIAEE, Raleigh, North Carolina, USA, pp. 560-569.

—. (2006). *Factors affecting agricultural extension personnel's motivation level.* Paper presented at the 22nd Annual Conference of AIAEE "International Teamwork in Agricultural and Extension Education", May 14 to 19 Clearwater beach, Florida.

Sandmann, L. R., & Vandenberg, L. (1995). A framework for 21st century leadership. *Journal of Extension, 33*(6).

Sasidhar, P. V. K., Rao, B. S., & Sreeramulu, P. (2008). Factors of role conflict among livestock extension professionals in Andhra Pradesh, India. *The Journal of Agricultural Education and Extension, 14*(4), 319-327. doi: 10.1080/13892240802416228

Schwarz, M. H., & Jerry D, G. (2006). *A need assessment of agricultural extension agents, specialists and program administrators in extension programming.* Paper presented at the 22nd Annual Conference of AIAEE "International Teamwork in Agricultural and Extension Education", Clearwater beach, Florida, pp. 617-626.

Shirzad, H. (2006). Leadership in extension organizations Retrieved 1, December, 2010, from http://neoruralism.blogspot.com/

Sulaiman, R., & Ban, A. W. V. D. (2003). Funding and delivering agricultural extension in India. *Journal of International Agricultural and Extension Education, 10*(1), 21-30.

Thach, L. N., & Ismail, M. (2010). *Motivation and extension skills as predictors of extension agents' performance in the Mekong Delta, Vietnam.* Paper presented at the International Conference on Developing Tomorrow's Leaders in Agricultural and Rural Development, Southeast Asian Regional Center for Graduate Study and Research in Agriculture (SEARCA), Los, Bonos, Laguna, Philippines.

Thompson, T. S. (2002). *A History of Extension at USAID* Paper presented at the "Extension and Rural Development: A Convergence of Views on International Approaches?" , Published by the World Bank and the United States Agency for International Development,, Washington, DC.

VanDenBan, A. W. (2009). Agricultural Extension: Worldwide Innovations (book review). *Journal of Agricultural Education and Extension, 15*(1), 107-114.

Vijayaragavan, K., & Ghosh, S. (2006). *Development oriented performance appraisal system to enhance productivity of extension professional.* Paper presented at the 22nd Annual Conference of AIAEE "International Teamwork in Agricultural and Extension Education", Clearwater Beach, Florida.

Vijayaragavan, K., Singh, P., & Wason, M. (2005). *Developing training modules for improving management skills of extension professionals.* Paper presented at the 21st Annual Conference" Educational, Extension and Research Strategies for a Changing World", San Antonio, TX, pp. 13-24.

Zhou, Y. (2009). Reinventing agricultural extension to smallholders www.syngentafoundation.org/view/element_href.cfm?src=1/524.pdf

CHAPTER TWO

ACHIEVEMENT MOTIVATION IN THE LEADERSHIP ROLE

This chapter covers the historical background of motivation theories and that of achievement motivation. It is then followed by the two theories related to achievement motivation that have been selected, after reviewing the achievement motivation theories and models, to explain the concepts and factors influencing achievement motivation in the leadership role of extension agents. The theories are Bandura's social cognitive theory (2001) and Farmer's achievement motivation model (1997). Bandura's theory tries to recognize the thinking processes that individuals go through in decision making for their behaviour by considering the reciprocal relationship of environmental factors on a person's cognition. Farmer's model expanded Bandura's theory by considering the pre-influence of background factors. The chapter ends by outlining the three dimensions of achievement motivation, namely aspiration, mastery, and salience.

Historical Background of Motivation Theories

The initial approaches to understanding human motivation were directed by Greek philosophers. In that time philosophers believed that the centre of individual efforts was focused on looking for enjoyment and keeping away from pain. The subject of motivation migrated from the area of philosophy to the newly rising psychological sciences around the end of the nineteenth century. Towards the 1930s, social scientists considered the human motivation and impact of social setting on behaviour (Steers, Mowday, & Shapiro, 2004). It was at nearly the same time that motivation theories played a vital role in extension education context, when extension executives considered promotion for encouraging rural people to implement their advice (Malek Mohammadi, 2000). Thus, from the early stage of development of motivation theories, extension education was a foundation for implementation and development of such theories, because extension is engaged with motivating others to change behaviour, similar to the purpose of motivation theories.

Oberhauser, Egea, Gilbert and Lopez (2004) believed that, thousands of years before the subject of motivation was inscribed to the research terminology, communities understood the significance of "carrot-and-stick motivation", which means encouraging workers to complete duties for an institute. In 1933, Elton Mayo, as the earliest study in motivation, showed the insufficiency of financial incentives or the pure carrot-and-stick motivation analogy.

Oberhauser et al. (2004) further categorized motivation theory in content and process theories. Content theories revolve around the recognition of inward needs, while process theories revolve principally around why individuals act in a particular manner, which is mostly related to factors such as perception and learning. The early theories of motivation were mostly categorized as content theories. While process theories as an innovative approach to the study of motivation appeared in the 1960s (Oberhauser et al., 2004). The end of the 1960s and the beginning of 1970s were called the "golden age" of motivation theories because of a huge movement on theory generation regarding influencing factors on motivation (Steers et al., 2004).

Content theorists concentrate on classification of the common needs with definite categories. Needs consists of primary needs, such as need for food which is physiological in nature, and generally inborn and secondary needs, such as need for achievement. In this category Maslow and Alderfer concentrate on the inner needs, while Herzberg mostly concentrates on hygiene and motivational factors in job situation. Meanwhile, McGregor's X and Y theory (McGregor, 2002) is a content theory in which X is based on workers caught in the lower levels (1 to 3) of Maslow's theory, similar to Herzberg's Hygiene, and Y is for workers who have gone above level 3 (Clark, 2008).

McClelland's theory (1985) explains the nature of three motivational needs in which need for achievement is associated with interest and surprise, need for power is associated with anger and excitement, and need for affiliation is associated with joy, happiness, and pleasure. According to McClelland, each primary motive is associated with positive emotions, and negative emotions may be associated with the avoidance motives. While the latter was not well elaborated in the McClelland theory, it was tested by later researchers such as Zurbriggen and Sturman (2002), who found significant results. The initial leadership motives according to McClelland are high need for power and activity inhibition, and low affiliation and achievement (McClelland & Boyatzis, 1982), and there are no sex differences in motivational predictors of attained leadership level (Jacobs & McClelland, 1994). Also the initial experiment of McClelland about achievement motive was to ask participants to throw rings over a peg from any distance they choose, in which achievement-motivated people threw the rings not so close as to create a ridiculously easy task, and not too far away to create an impossible task. They arranged fairly difficult but achievable targets (McClelland, 1971), which is a criterion of mastery of motivation, which, according to Farmer's model, is just one dimension of achievement motivation. Although the McClelland theory was an important contributor in understanding psychological aspects of self-attributed and implicit social motives partly was related to unconscious networks of cognitive process (McClelland, Koestnera, & Weinbergera, 1989), and more specifically, achievement motivation (Oberhauser et al., 2004), it mostly explained the psychological nature of motivational needs (Elliot & Dweck, 2005). While due to the specific criteria of extension systems, extension agents, through learning and teaching process, play an emergent leadership role to the clients, in which a more social-based learning theory is demanded to explain the concept and influencing factors.

Process theories, in comparison with content theories, focus on distinguishing factors linked with motivation in a comparatively static situation. The process theorists' point of view is to see motivation in a dynamic perspective and seek for causal relations across time and events while they are connected to human behaviour in the workplace. The main process theories are cognitive motivation theories that communally try to recognize the thinking processes that individuals go through in decision making for their behaviour in the workplace, such as Bandura's theory (Steers et al., 2004), which will be discussed later. Among process theories of motivation, Vroom's expectancy theory refers to motivation as the product of valence (reward), expectancy (performance), and instrumentality (belief) (Recklies, 2001), which is all about individual motivation to achieve. The expectancy theory discusses individuals' decision making processes and allocation of attempts, which is also a key aspect considered in social learning theory. Further, the Vroom expectancy theory is based on the research of Lewin and Tolman, who proposed their theory of "resultant valence" to account for aspiration processes (Steers et al., 2004). Also there is a neglected point about the importance, and the effect of context or environmental factor, in Vroom's theory. Lastly, this theory considers reward as the motivator, while in Herzberg it is the hygiene factor that is more tangible. Thus, one could conclude from content and process theories that process theories are more suitable to investigate factors influencing achievement motivation of extension agents. Also among process theories, as will be discussed in the following section, Bandura's social cognitive theory is selected. The next section more specifically discusses the background of achievement motivation theories.

Historical Background of Achievement Motivation Theories

Research on achievement motivation also has a long and notable history. Achievement motivation was considered as a scientific discipline when James, in 1980, offered speculation about how achievement striving is linked to self-evaluation. However, truly programmatic empirical work on achievement motivation began earlier in Kurt Lewin's laboratory in 1938 with the investigation of aspiration-setting behaviour. Lewin and colleagues proposed their theory of "resultant valence" to account for aspiration processes (Elliot & Dweck, 2005). Later, around the 1960s, the central place of research on achievement motivation in scientific psychology was solidified by the work of McClelland, Atkinson, and colleagues, on need for achievement (Elliot & Dweck, 2005). At the same time, social learning theory launched and prepared a basis for social cognitive theory in 1986 (Stone, 1998). Indeed, researchers made great advances in conceptual developments and experiential research regarding achievement motivation, of which the three main developed conceptualizations will be explained in the following paragraphs.

Balu and Dancan's model, as well as Wisconsin's model, were two of the initial models in the achievement motivation area around the 1960s. Balu and Dancan's model concentrates on the structure of status transmission by considering the father's educational achievements and occupational prestige. While the main focus of the Wisconsin model is the social psychological dynamics mediating interpersonal influences on the individual attainment by considering a person's mental ability, peer influence, and personal aspiration. Most of the researchers in the status attainment area have discussed the effect of the family socioeconomic status (and school location or support) as the most significant predictors of educational attainment and then occupational attainment (Haller & Portes, 1973). Research on these models and status attainment theory has mostly concentrated on predicting occupational achievement motivation among young subjects by considering the family socioeconomic status.

London and Noe (1997) were among the more recent contributors in achievement motivation. Their model, as shown in Figure 2-1, was implemented as a framework for conceptualizing elements of motivation, studying their antecedents and consequences, and designing interventions to enhance career success and job performance. They integrate individual differences into three domains for determining the concept of career motivation: career resilience, career insight, and career identity. The model was originally developed in the process of answering a research question, examining what motivates managers to attain a leadership role (London & Noe, 1997). This model with a philosophical view worked on assigned leadership role of managers, and differs from the current analysis that focuses on emergent leadership role.

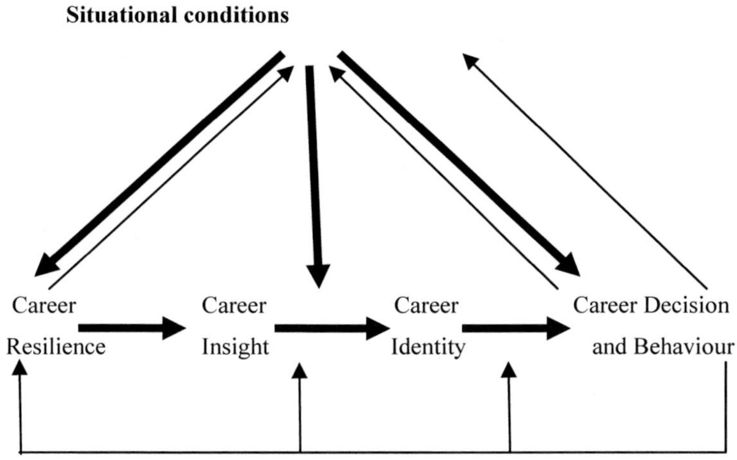

Figure 2-1: The model of career motivation of London and Noe (1997; p.62)

Bandura's Social Cognitive Theory (SCT)

The Social Cognitive Theory (SCT) has a rich historical background and stems from the Social Learning Theory (SLT). Albert Bandura first published SLT in the 1960s, and formally launched the SCT (see Figure 2-2) in 1986 within his book entitled *Social Foundations of Thought and Action: A Social Cognitive Theory* (Stone, 1998).

The SLT focuses on the observing and modelling the behaviours, attitudes, and emotional reactions of others. It emphasizes learning through observation and modelling. The theory was developed initially based on a behaviourism perspective but at present embraces many cognitivists' thoughts. Since the SCT supports the behaviourist view, stating that response consequences mediate behaviour, it asserted that behaviour is mainly regulated antecedently through cognitive processes. Hence, people give meaning to the events and are inclined to feel positively or negatively based on the meaning of each event (Bandura, 2000).

SLT discusses the role of cognitive and environmental factors and how they interact to control individuals' learning and behaviour. This theory argued that learning takes place in a social framework and individuals learn from each other, through observation, imitation, and modelling (Bandura, 1999). Bandura declared that individuals attain behavioural changes in answer to environmental factors and try to sustain the changes. He believed that people are not passive reactors to the environment, people actively shape their environments and it is the best place to see human nature, rather than focusing on rewarded behaviours (Bandura, 2001). SCT was the initial theory to integrate the concept of modelling, or vicarious learning in social learning. Bandura also introduced some other key concepts, including reciprocal determinism and self-efficacy (Bandura, 1989).

Self-efficacy refers to the learners' self-confidence towards learning. The likelihood of getting involved in specific behaviour is higher when people are supposed to be competent in implementing that behaviour effectively (Bandura, 2006). Bandura's reciprocal causation exists because behaviour can influence both the environment and the person (Amarasing, 2008), although they may not be of equal strength (Alderman, 2004). In this view the person's cognitive processes, such as belief about his or her ability, act as an important mediator of motivation which influences both expectations for future performance and action taken. Presently researchers subscribe to variant forms of the SLT which all share three fundamental beliefs as follows:

First: Response consequences such as rewards or punishments which affect the probability that a person will carry out a particular behaviour again in a given situation. This tenet is similar to classical behaviourists.

Second: Individuals can learn by observing others, in addition to learning by doing. Learning by observing others is called vicarious learning. This tenet of vicarious learning is not similar to classical behaviourists.

Third: Individuals are most likely to model the observed behaviour of others they identify with. Identification with others is the degree to which a person is perceived to be similar to one's self, in addition to the degree of emotional attachment that is felt toward an individual (Stone, 1998). Two types of model exist; they are the live model, which is about real individual representing the behaviour; and the symbolic model, in which an individual or character is portrayed in media or a book (Amarasing, 2008).

Bandura's social cognitive theory and concept of self-efficacy received vast attention from different scholars, such as those who are interested in organizational and career behaviour, who then provide a framework to assist in promotion of adaptive motivation strategies. In such a piece of research, Lent (2007) introduced Bandura's theory as his foundation for the social cognitive career theory (SCCT), which is one of the popular theories with implications for expanding vocational interests, clarifying career goals, instilling realistic outcome expectations, and building environmental support systems. SCCT discusses the basic concept of self-efficacy, outcome expectation, goal and contextual supports and barriers.

Bandura's social cognitive theory is selected as the underlying theory for explaining factors influencing achievement motivation in the leadership role of extension agents due to its importance in understanding motivation at the individual level. Also this theory has the strengths of behavioural consideration on the effect of environment along with cognitive strengths by trying to recognize the thinking processes that individuals experience in determining how to behave in the workplace. In addition, Bandura's social cognitive theory, as one of the well-known process theories of motivation, could explain the factors associated with motivation and see motivation as a dynamic process relating events and behaviour at the workplace of employees (Steers et al., 2004). Bandura's theory is the best in the context of extension since extension agents play a modelling role and try to lead clients through observational learning, which is a basic premise of this theory.

Thus Bandura's theory is used to explain the achievement motivation in the leadership role of extension agents, which covers the mediation effects of two main factors, namely personal and environmental. Bandura's social cognitive theory has the potential of downward theory building, specifically in the context of extension education, if we could carry out interpretive research frequently in the form of pilot studies to make grand range theories, such as SCT, more testable (Williams, 2003).

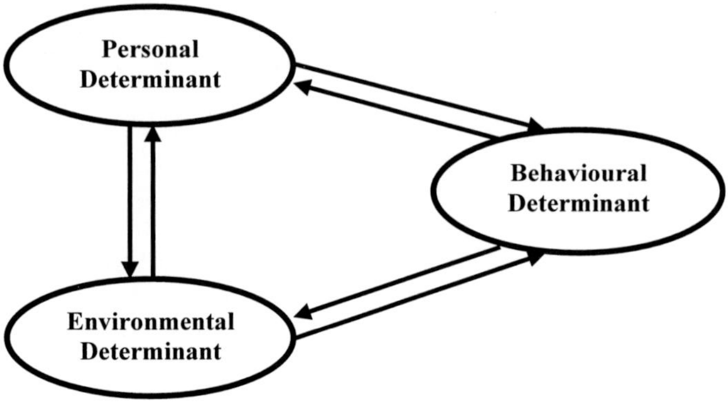

Figure 2-2: Social cognitive model of Bandura (2001)

Farmer's Model of Achievement Motivation

Farmer (1985; 1997) developed her model during longitudinal studies. Figure 2-3 shows the basic model of Farmer's studies, including some of the variables of her longitudinal studies. Farmer's multidimensional model discusses three sets of factors, including background, personal, and environmental factors, which have influence on three motivation dimensions, namely aspiration, mastery, and salience. Farmer's dimensions of achievement motivation include aspiration, mastery, and salience that are parallel with career identity, insight, and resilience, as in the conceptualization of career motivation by London and Noe (1997). Farmer's model permits rising motivation by targeting change efforts on a number of the changeable environment and personal factors. The model also allows for focusing on dimensions of achievement motivation among men and women, while other models and theories are silent about the gender perspective.

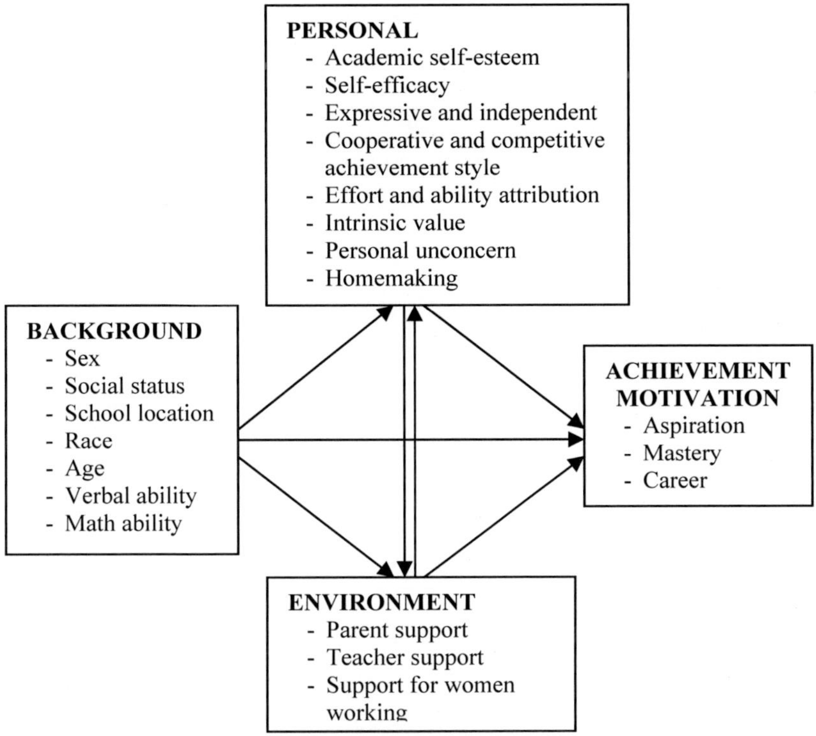

Figure 2-3: The achievement motivation model of Farmer (1985, 1997; p.365)

Farmer's (1997) model of career achievement motivation is grounded in social learning theory and social cognitive theory. The model was used in the design of a longitudinal study from 1978 and in 1980, 1990, 1991, and 1993 with young male and female subjects. This model emphasized how background factors have pre-influence on personal and environmental factors in career achievement motivation. Farmer's (1997) model is used in this book to explain the concept of achievement motivation using the three dimensions of aspiration, mastery, and salience, which will be discussed in the following sections. Additionally, Farmer's (1997) model will be used to explain how personal and environmental motivators mediate the effect of background factors on achievement motivation of extension agents in leadership role, which will be presented in the next chapter.

Motivation and Achievement Motivation

The term motivation is obtained from a Latin word for movement (movere). Atkinson defined motivation as the contemporary (immediate) influence on direction, vigour, and persistence of action, while Vroom defined motivation as a process governing choice made by persons among alternative forms of voluntary activity (Steers et al., 2004).

Through the years, motivation was considered to have three psychological functions. First, energizing or activating behaviour, which means what gets individuals engaged in or turned off toward learning; second, directing behaviour, which considers why one course of action is chosen over another; third, regulating persistence of behaviour, which refers to why individuals persist toward goals (Alderman, 2004). Further, the three constituents of motivation were considered as direction, effort, and persistence. Direction was defined as what an individual is attempting to do. Effort was related to individuals' hard work, and persistence was related to individuals continuing to try (Armstrong, 2006). One could argue that these three components of motivation are parallel with the three dimensions of achievement motivation explained by Farmer (1997), which include aspiration, mastery and salience.

According to Fabusoro, Awotunde, Sodiya, and Alarima (2008), motivation is defined as a process through a forceful need that stimulates and controls interest in achieving expected targets. Individuals' motivation is a psychological process that includes setting goals, guiding, and strengthening of feeling. Motive of actions would be a pressing physical or mental need or desire that is not required for survival; however, it is important for achieving arranged targets.

Achievement motivation is defined as a tendency to try hard in the direction of performance excellence (Cox, 2008). Researchers who study achievement motivation are interested in why individuals are engaged in the variety of achievement related behaviours. Achievement motivation theorists try to clarify an individual's choice of achievement tasks, persistence with those tasks, vigour in carrying them out, and performance in them (Wigfield & Eccles, 2000).

The motivation knowledge has been expanded from formal theories such as social learning. Although the majority of the recent approaches similarly focus on beliefs and cognitive and emotional variables that influence achievement effort (Alderman, 2004), literature on achievement motivation is weak in two fundamental aspects. First, definition of the achievement concept, and second, understanding of how achievement should be conceptualized (Elliot & Dweck, 2005). Elliot and Dweck

(2005) believed in the need for competence as a primary motivation that assists individuals to adapt to the environment. Thus, they proposed competence as the conceptual centre of achievement motivation that could show the weakness of achievement motivation studies. The motivational analysis of competence could explain the ways in which individual behaviours are energized and directed. Thus, the present book concentrates on leadership role, proposed as a foundation of a key competency of extension workers in conceptualizing achievement motivation.

As was alluded to earlier, Farmer (1997) defined achievement motivation through three dimensions: aspiration, mastery, and salience. Bandura's social cognitive theory concept of self-regulatory systems could support the definition of achievement motivation through three motioned dimensions. The self-regulatory systems permits people to control their own thoughts, feelings, motivations, and actions, formed through reciprocal influence of self and outer sources, which consist of social and moral standards of motivation. The motivational standards guided the behaviour in a process of "discrepancy production" (goal setting) and "discrepancy reduction" (work to attain a goal). The process of goal arrangement and then comparison of goal with standards and personal achievement could motivate a person to work harder or change their behaviour (Stone, 1998). One could argue that the desire for goal setting is near to aspiration motive, and working hard to attain a goal is consistent with mastery and commitment (salience) motive. It is in agreement with London and Noe's (1997) study, which considered desire goal setting as an indicator of situational characteristics affecting aspiration, and hard work to attain a goal as the supporter of mastery and salience motives.

Achievement motivation of leadership role in extension context differs from other contexts due to the way extension agents are responsible for behaving with clients, and the expectations and characters of the latter. It also refers to initiatives or efforts made by extension agents to lead farmers and their families. More specifically, achievement motivation in leadership role of extension agents is a desire for attainment of excellence in the leadership role for achieving improved performance, including the three dimensions of aspiration, mastery, and salience, which will be described in later parts of this volume.

Aspiration

Aspiration refers to "the level of education and occupation to which a person aspires" (Farmer, 1985, pp. 366). Aspiration for the workforce in the workplace, according to Ghosh and Vijayaragavan (2003), is to plan and work for achieving improved performance, while career aspiration is an individual's orientation toward a particular career goal (Domenico & Jones, 2007). Aspiration as a dimension of achievement motivation was supported in the literature in addition to Bandura's social cognitive theory, through status attainment theory, which discussed a correlation between an individual's background (e.g. parental social status) and cognitive ability, along with other environmental factors and their impact on desired levels of educational status and then career attainment (Farmer, 1985; Ottino, 2009). Thus, status attainment theory discussed achievement motivation through the aspiration and desired level of attainment, although in this book status attainment or achievement does not mean attainment of status in hierarchy, but rather attainment of excellent performance in extension context. Thus, the desired level of attainment of excellence in current status regarding the leadership role is the meaning of aspiration as a dimension of achievement motivation. In other words, when the achievement motivation is defined as the attainment of excellence, in this case, leadership role, then the aspiration is considered as the desired level of leadership attainment.

Career aspiration is a dimension of internal career that fundamentally emerges from an individual, which in turn influences one's values, norms, and beliefs (Ramly, Ismail, & Uli, 2009). Career aspiration pursues a particular career path, and is regarded as a selected goal, which is an intention to act in the near future (Larson, Wei, Wu, Borgen, & Bailey, 2007). Socialization of gender role in each socio-environmental context shapes special career aspiration for the members of each sex (Powell & Graves, 2003). It is a possible explanation for the research findings such as those of Farmer (1985; 1997) who considered background factors as the first factors affecting achievement motivation due to crystallization of aspiration in an early stage, under the influence of gender and gender role, as well as some other variables. London and Noe (1997) further articulated that aspiration, which is considered the first construct of achievement motivation, affects mastery and then these two affect resilience, such as sequence. In the case of aspiration as a dimension of achievement motivation, the order of factor effects is background, environmental, and then personal factors, which are different from mastery and salience dimensions. In mastery and salience, after the background factor, personal, and then environmental factors contribute to achievement motivation (Farmer, 1985). Thus, aspiration as a first construct of achievement motivation that affects the other construct (Manuel London & Noe, 1997) was mostly influenced by background factors (Farmer, 1985).

Although researchers such as Domenico and Jones (2006) believed aspiration was crystallized during the early stage of a person's life, according to the aspiration adaptation theory of Sauermann and Selten that goes back to 1962, aspiration is adjustable. This theory models decision making in a multi-goal framework where each goal has a different aspiration level (Gigerenzer & Selten, 2001). According to the above mentioned discussion, aspiration is adjustable and may go up or down depending on different job goals and different situations. The important point from this is that investigation of influencing factors on adjustable aspiration is more significant in comparison with crystallized aspiration. Thus, considering extension agents who have different roles, composed of different career goals that have to be well practiced for a well performing extension job, aspiration levels in performing each career goal also would be different, making investigation of factors influencing their aspiration important to maximize the performance and job outcomes.

Further, according to aspiration adaptation theory, aspiration is magnitude of goal worth. The worth or value of each goal, which is equivalent to aspiration level, varies in separate steps. The adjustable aspiration levels of goals shape an aspiration grid. The present aspiration level for modifying a goal shifts to a nearest up or down point on the aspiration grid, which respectively showed increase or decrease of a goal's aspiration. The assignment of an urgency order, along with aspiration grid, shapes an aspiration scheme. The aspiration scheme in aspiration adaptation theory takes the place of preference order in the common decision theory (Gigerenzer & Selten, 2001).

Gender was confirmed over the past thirty years by researchers as one of the important factors influencing aspiration (Cannon, 2004). Domenico and Jones (2006) believe that the reason for studying women's career aspirations is their contribution to the workforce. Further, women's career aspirations were mostly influenced by socioeconomic status, race, parents' occupations, educational level, and parental expectations. The social status as a factor influencing aspiration could be interpreted as occupational position, where each position brings special status and could affect individual aspiration.

Hensley (2002) argued that individuals with androgynous or masculine gender role have higher career aspiration than feminine individuals and he only found a significant relationship for male subjects. Also Karami, Ismail, and Md.Sail (2011) showed that masculinity had a significant positive effect on respondents' aspirations. This is consistent with Powell and Butterfield's (2003) study results, which showed men were more likely than women to aspire to top management and a gender identity consisting of high masculinity, and low femininity was associated with higher aspirations. Powell and Butterfield's (2003) research also found similar results to those of Chugh and Sahgal (2007).

Therefore, aspiration in the leadership role of extension agents is defined operationally as the desired level of a job role, which is leadership role. Also, it is assumed that partial aspiration levels through adjustment steps would be varied in different personal and environmental situations (Gigerenzer & Selten, 2001). Thus, different aspiration levels could increase or decrease the achievement motivation and consequently performance level of extension agents according to aspiration adaptation theory. Therefore, an investigation of influencing factors on aspiration that is adjustable would help extension organizations to improve the extension agents' aspirations.

Mastery

Mastery motivation refers to "the tendency of persons to choose difficult challenging tasks rather than easy tasks and to keep struggling to master the task once they have started" (Farmer, 1985, pp. 366). This dimension of achievement theory, in addition to Bandura's theory (Leal, 2006), in the literature was supported by early work of McClelland and Atkinson whose theory was referred to as achievement motivation (Farmer, 1985; Ottino, 2009).

McClelland developed achievement-based motivational theory and improved worker appraisal methods based on competency, better than conventional intelligence quotient (IQ) and personality-based tests (Kerr, 2009). The initial McClelland experiment was to ask participants to throw rings over a peg from any distance they chose. The majority of individuals threw at close or far away random distance; but achievement motivated people threw the rings not so close as to create a ridiculously easy task, and not too far away to create an impossible task. They arranged fairly difficult but achievable targets (McClelland, 1971). McClelland believed that achievement motivated people arranged targets that they can achieve through attempts and skill, and they appear to be more concerned with personal achievement rather than rewards of accomplishment. He revealed individuals who have high achievement need are more interested in new ideas (Kerr, 2009). Achievement motivated people considered rewards such as money just as a measurement of their performance compare with others (McClelland, 1971). Achievement motivated people spend time to think how they can do the tasks better, and they consider feedback to which they answered positively, and utilize this knowledge to do the task better (Kerr, 2009). Thus, McClelland's theory discussed achievement motivation through mastery motivation and desire to choose difficult, challenging tasks, but accessible along with interest for personal achievement rather than the rewards of accomplishment and request for related feedback.

The concept of mastery motivation has been examined by several writers, of which Dweck's 1986 and 1991 research are the most well-known. Considerable mastery motivation orientations are characterized by more concern with achieving success, rather than with avoiding failure, although failure is not simply taken to imply a lack of ability that precludes future success. Thus, it is respected that the mastery oriented ought to be more able to display task orientation and incremental views concerning the nature of ability (Daniels & Edwards, 2004).

Mastery motivation was most affected, according to Farmer (1985), by personal influences and, to a lesser extent, by background and environmental influences. The fact that mastery motivation is affected more by changeable personal influences than unchangeable background influences suggest that mastery motivation may change over a person's lifetime, since younger subjects had higher mastery motivation than older students. Patel (2008) longitudinally investigated changes in children's achievement goals over the course of first-grade, with a specific focus on the role of temperament and the school context. The results revealed that multiple temperament patterns and the academic environment interacted to form children's achievement goals.

Thus, mastery as a dimension of achievement motivation in the leadership role of extension agents is defined operationally as desire for intellectual challenge, choosing difficult, challenging, but achievable tasks, and continuing to struggle to master the task in the leadership role.

Salience

Salience refers to "the extent to which a person sees involvement in a career as central to his or her adult life role" (Farmer, 1985, pp. 366). Sweeney (2004) defined career salience as the value people put on the work role, which can be modified during their life. Super and Nevill (1984) defined work salience as the relative importance of work for individuals over a long span, which, as interests and opportunities change, it will also change over a life-span. Jong (1999) defined career commitment, which is a paralleling concept with salience, as the degree to which contribution in a job is central for a person in the present and future.

Since Farmer (1985; 1997) considers achievement motivation as a kind of career development, she supported her idea about counting salience as a dimension of achievement motivation using Super's theory. Super's life-span, life-space conceptual framework defined a career as the number of roles performed by individuals during lifetime. Furthermore, he stated that career development evolved in life-space, which is the social context, during the course of a lifetime, which refers to life-span (Ferreira, Santos, Fonseca, & Haase, 2007; Super, 1980). The life space concept in Super's theory acknowledged that individuals attach different degrees of importance to jobs and job roles. The attachment of salience people apply to the constellation of life roles that they play, defines their life structure (Super & Nevill, 1984). Super and Nevill (1984) revealed that the concept of salience is an important matter in career development. They also showed that work salience is slightly affected by gender under the effect of role conflict, and it is not related to socioeconomic status.

According to London and Noe (1997) employees with high levels of career salience may be able to cope better with change than people with low levels of salience. Employee salience is important in the current situation of organizational restructuring and downsizing, which are common phenomena. Organizational restructuring and downsizing may cause life stresses and consequently may cause people to question the significance of their goals and their level of capabilities, in turn, leading to career shifts. Some people never overcome career stress, while others are able to overcome self-doubt and commit themselves to career and life goals (London & Noe, 1997).

Although there are studies that showed gender differences in terms of career commitment or salience, such as Farmer's (1997) studies, the results of Sweeney's (2004) study among men and women head coaching showed no gender differences for career salience. In my point of view, gender difference exists not because of biological sex but is mostly related to the pre-influence of different background factor variables that were shaped under the effect of some stereotypical assumption, which consequently affect the personal and environmental factors.

Ladebo, Olaoye, and Adamu (2008) mentioned that the literature on the work attitudes of extension workers, such as satisfaction and organizational commitment, is not very developed. The majority of studies in the area of extension agents' work attitude investigated the demographic characteristics of extension personnel, such as age, gender, race, marital status, and education. The authors studied extension personnel from Ogun State in Southwest Nigeria. They hypothesized that extension agents' commitment includes commitment to different key actors of an organization, such as the organization itself, a supervisor, and/or workgroup members. They found positive relations between self-esteem and organizational commitment ($\beta = 0.58$, $p<0.0001$), supervisor commitment ($\beta = 0.28$, $p<0.0001$), and workgroup commitment ($\beta = 0.33$, $p<0.0001$).

Jong (1999) found a positive relationship between intrinsic and extrinsic motivation with career commitment. Although career commitment of women was mainly predicted by intrinsic motivation, male career commitment was predicted through both intrinsic and extrinsic motivations.

Therefore, salience in the leadership role of extension agents is considered operationally as the extent to which a person sees involvement in leadership role as central to his or her adult life role. Hence, the achievement motivation and its constructs (aspiration, mastery, and salience) have been discussed in this chapter. The following chapters discuss factors influencing achievement motivation that consist of background, personal, and environmental factors.

Conclusion

The reviewed theories of motivation provide a historical perspective of the achievement motivation theories (see Fiqure 2-4), in which process theories are preferred due to their focus on motivation in a dynamic perspective and that they seek for causal relations across time and events. From the process theories and models, Bandura's social cognitive theory and Farmer's model of achievement motivation give priority to explaining the factors affecting achievement motivation in the leadership role of extension agents. Bandura's theory tries to recognize the thinking processes that individuals go through in decision making for their behaviour by considering the reciprocal relationship of environmental factors on a person's cognition. Farmer's model also extended Bandura's theory by considering the pre-influence of background on achievement motivation. All three factors, including pre-influence of background factors along with mediation effects of personal and environmental factors and their underlying variables, will be discussed in the next chapters.

Further, investigation of motivation and achievement motivation research and theories showed that most of the researchers and theorizers consider only one dimension of achievement motivation in defining and measuring achievement motivation, while according to Farmer (1985; 1997) aspiration, mastery and salience all cause achievement motivation, which had been discussed in this chapter. Thus, taking a multidimensional look at achievement motivation, and giving attention to the pre-influence of background factors on achievement motivation according to Farmer's model, would make a substantial stride beyond Bandura's social cognitive theory. In spite of there being a lack of theory on emergent leadership role achievement motivation, the aforementioned discussion could fill the gap in the body of knowledge pertaining to achievement motivation in the workplace, with a main focus on emergent leadership role of extension agents in the extension context.

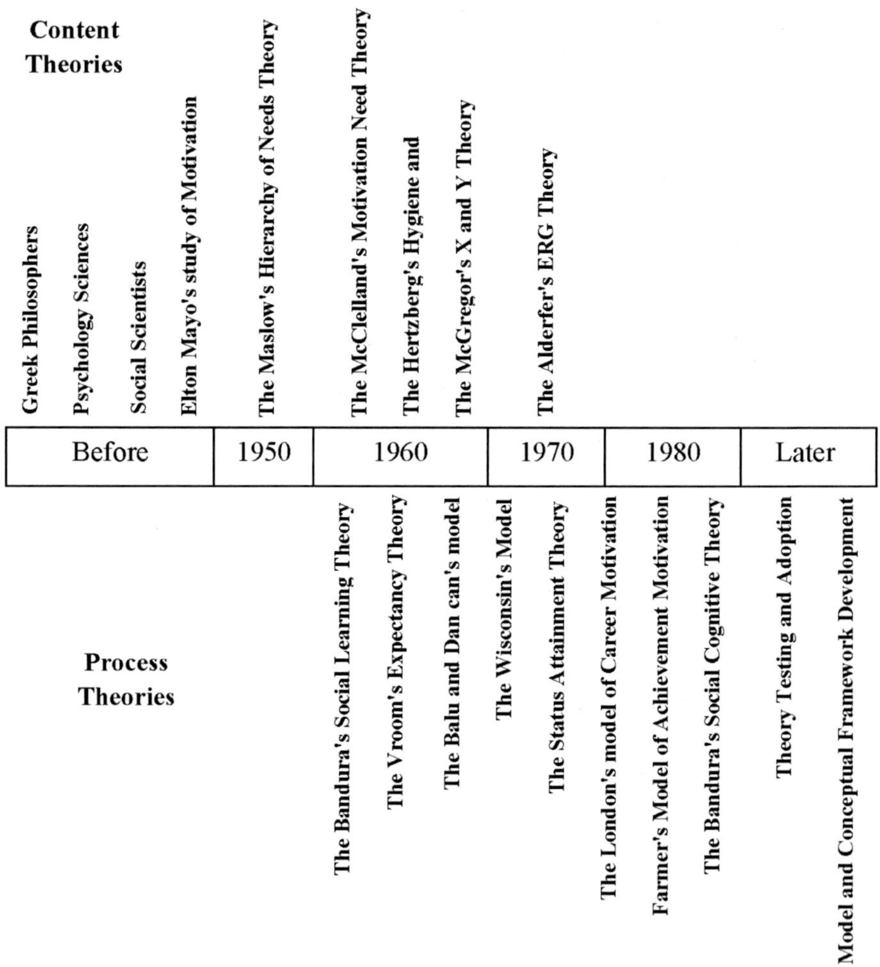

Figure 2-4: The historical diagram of motivation theories

References

Alderman, M. K. (2004). *Motivation for achievement: possibilities for teaching and learning*. New Jersey: Lawrence Erlbaum Associates, Inc.

Amarasing, P. K. (2008). Social learning theory. Retrieved 8, Jan, 2009, from Online Learning Laboratory; University of South Alabama *www.usaoll.org/mobile/theory.../social_learning_theory.htm*

Armstrong, M. (2006). *A handbook of human resource management practice 10th edition.* London: Cambridge University Press.

Bandura, A. (1989). Social cognitive theory. *Annals of child development, 6*, 1-60.

—. (1999). Social cognitive theory: An agentic perspective. *Asian Journal of Social Psychology 2*, 21-41.

—. (2000). Cultivate self-efficacy for personal and organizational effectiveness. In E. A. Locke (Ed.), *Handbook of principles of organization behavior*. Oxford, UK: Blackwell.

—. (2001). Social cognitive theory of mass communication. *Media Psychology, 3*(3), 265-299.

—. (2006). Guide for constructing a self-efficacy scales. In F. Pajares, & Urdan, T. C. (Ed.), *Self-efficacy beliefs of adolescents.* United State of America: IAP-Information Age Publishing ,Inc.

Cannon, S. C. (2004). *Supporting first-year college women in math, science, and related majors: A Career Development Intervention (Doctoral dissertation).* Doctor of philosophy, North Carolina State University, United State.

Chugh, S., & Sahgal, P. (2007). Why do few women advance to leadership positions? *Global Business Review, 8*(2), 351-365.

Clark, D. (2008). Leadership & human behavior. Retrieved 10 Jun, 2009, from http://nwlink.com

Cox, K. S. (2008). *Motivational factors influencing women's decision to pursue upper-level administrative position in higher education (doctoral dissertation).* PhD, Kansas State University Manhattan, Kansas.

Daniels, H., & Edwards, A. (2004). *The routledgefalmer reader in psychology of education* London: Taylor and Francis group.

Domenico, M. D., & Jones, H. K. (2006). Career aspirations of women in the 20th century. *Journal of Career and Technical Education, 22*(2), 18-25.

—. (2007). Career aspiration of pregnant and parenting adolescent. *Journal of Family and Consumer Sciences Education, 25*(1), 24-33.

Elliot, A. J., & Dweck, C. S. (2005). *Handbook of competence and motivation*. New York: Guilford press.

Fabusoro, E., Awotunde, J. A., Sodiya, C. I., & Alarima, C. I. (2008). Status of job motivation and job performance of field level extension agents in Ogun State: Implications for agricultural development. *Journal of Agricultural Education and Extension, 14*(2), 139-152.

Farmer, H. S. (1985). Model of career and achievement motivation for women and men. *Journal of Counseling Psychology, 32*(3), 363-390.

—. (1997). Women's motivation related to mastery, career salience, and career aspiration: A multivariate model focusing on the effects of sex role socialization. *Journal of Career Assessment, 5*(2), 355-381.

Ferreira, J. A., Santos, E. J. R., Fonseca, A. n. C., & Haase, R. F. (2007). Early predictors of career development: A 10-year follow-up study. *Journal of Vocational Behavior, 70,* 61-77.

Ghosh, S., & Vijayaragavan, K. (2003). Performance appraisal climate of organizations: An influencing factor for overall job satisfaction of extension personnel. *The Journal of Agricultural Education and Extension, 9*(1), 1-9. doi: 10.1080/13892240385300031

Gigerenzer, G., & Selten, R. (2001). *Bounded rationality: the adaptive toolbox*. Cambridge, Mass, USA: The MIT Press.

Haller, A. O., & Portes, A. (1973). Status attainment processes. *Sociology of Education 46*, 51-91.

Hensley, L. (2002). *The Influence of Gender Role and Gender Stereotyping on the Career Choice and Career Commitment of Adolescents.* Retrieved 29 Jan, 2009
http://www. samford.edu/schools/artsci/scs/hensley.html.

Jacobs, R. L., & McClelland, D. C. (1994). Moving up the corporate ladder: A longitudinal study of the leadership motive pattern and managerial success in women and men. *Consulting Psychology Journal: Practice and Research, 46*(1), 32-41.

Jong, E. d. (1999). The impact of motivation on the career commitment of Dutch literary translators from Elsevier Science
www.elsevier.nl/locate/poetic

Karami, R., Ismail, M., & Md.Sail, R. (2011). Gender identity and career aspiration to top management of Malaysian graduate students. *Asian Pacific Journal of Education, 31*(4), 439-454.

Kerr, B. (2009). *Encyclopedia of giftedness, creativity, and talent.* USA: SAGE publication, Inc.

Ladebo, O. J., Olaoye, O. J., & Adamu, C. O. (2008). Extension personnel's self-esteem and workplace relationships: Implications for job satisfaction and affective organizational commitment foci. *The Journal of Agricultural Education and Extension, 14*(3), 249-263.

Larson, M. L., Wei, M., Wu, T. F., Borgen, H. F., & Bailey, C. D. (2007). Discriminating among educational majors and career aspirations in Taiwanese undergraduates: the contribution of personality and self-efficacy. *Journal of Counseling Psychology, 54*(4), 395-408.

Leal, V. M. (2006). *Predictors of non-traditional career self-efficacy in Mexican American adolescent women (doctoral dissertation).* phd, The Ohio State University USA.

Lent, B. (2007). What attracts students to – and keeps them in – STEM Fields? A theoretical perspective Retrieved 11, May, 2009, from *www.dels.nas.edu/bls/more/Lent.pdf*

London, M., & Noe, R. A. (1997). London's career motivation theory: An update on measurement and research. *Journal of Career Assessment, 5*(1), 61-80.

MalekMohammadi, I. (2000). *Agricultural extension education and human resources*. Tehran, Iran: Tehran University Publication

McClelland, D. C. (1971). *Assessing human motivation*. Morristown, NJ: General Learning Press.

—. (1985). *Human motivation*. Cambridge, UK: Cambridge University Press.

McClelland, D. C., & Boyatzis, R. E. (1982). Leadership motive pattern and long-term success in management. *Journal of Applied Psychology, 67*(6), 737-743.

McClelland, D. C., Koestnera, R., & Weinbergera, J. (1989). How do self-attributed and implicit motives differ? *Psychological Review, 96*(4), 690-702.

McGregor, D. (2002). Theory X and theory Y. *Workforce, 81*(1), 32.

Oberhauser, C. L., Egea, M., Gilbert, S., & Lopez, A. (2004). Motivation theory. Retrieved 29 May, 2009, from *userwww.sfsu.edu/~gtarakji/engr801/.../motivation.html.*

Ottino, K. L. (2009). *Diminished aspiration: women central office administrators and the superintendency (doctoral dissertation).* PhD, The University of Minnesota United State.

Patel, D. M. (2008). *The effects of temperament and schooling on achievement motivation in first-grade children (doctoral dissertation).* Doctor of Philosophy George Mason University Northern Virginia.

Powell, G. N., & Butterfield, D. A. (2003). Gender, gender identity, and aspirations to top management. *Women in Management Review, 18*(1/2), 88-96.

Ramly, E. S., Ismail, M., & Uli, J. (2009). Antecedents of career aspiration of R&D professionals in Malaysian Public Organizations. *European Journal of Scientific Research, 26*(1), 66-79.

Recklies, D. (2001). Motivation – basic concepts and theories. Retrieved 13 July, 2009, from *www.themanager.org*

Steers, R. M., Mowday, R. T., & Shapiro, D. L. (2004). Introduction to special topic forum the future of work motivation theory. *Academy of Management Review, 29*(3), 379-387.

Stone, D. (1998). Social cognitive theory Retrieved 19, Sep, 2009, from *www.infosihat.gov.my/.../Social%20Cognitive%20Theory.doc*

Super, D. E. (1980). A life-span, life-space approach to career development. *Journal of Vocational Behavior, 16*, 282-298.

Super, D. E., & Nevill, D. D. (1984). Work role salience as a determinant of career maturity in high school students. *Journal of Vocational Behavior, 25*, 30-44

Sweeney, K. (2004). *The relationship between family commitment and career salience among coaches of elite NCAA division I-A women's programs (doctoral dissertation).* PhD, University of New Mexico, New Mexico.

Wigfield, A., & Eccles, J. S. (2000). Expectancy–value theory of achievement motivation. *Contemporary Educational Psychology, 25*, 68–81. doi: 10.1006/ceps.1999.1015.

Williams, M. (2003). *Making sense of social research.* London: SAGE Publication Ltd.

Zurbriggen, E. L., & Sturman, T. S. (2002). Linking motives and emotions: A test of McClelland's hypotheses. *Personality and Social Psychology Bulletin, 28*, 521-535.

CHAPTER THREE

BACKGROUND FACTORS AND ACHIEVEMENT MOTIVATION IN THE LEADERSHIP ROLE

A particular aspect that has remained neglected when investigating achievement motivation is considering the background factor which has pre-influence on personal and environmental factors (Powell & Graves, 2003). The characteristics that could shape background factor variables of leadership role, which is a role with a long stereotypical history, are most likely gender quality variables. The importance of gender quality is related to the gender ratio of extension agents and the importance of gender homophily of extension agents (as leaders) with clients. Meanwhile, more than 32 percent of the Iranian population live in rural areas and form clients to the extension agents. Women make up half of these clients, while the gender ratio of extension agents is only 9:1 men to women (Sadighi, 2003). The importance of leader homophily with clients is due to the norm in which women clients should be approached by women extension agent leaders. The problem becomes more serious when, in addition to gender ratio, some research results take into account the fact that women extension agents do not perceive themselves to be as strong as males in terms of leadership role (Navarro & Ricketts, 2008). A recent study of Jivetti and Edwards (2009) showed that self-help group women in Western Kenya perceived a lack of leadership competency as a factor affecting their performance. Research over the past thirty years according to Cannon (2004) has also demonstrated that two of the important factors affecting aspiration, which is a dimension of achievement motivation, are gender and gender identity. This chapter covers the conceptualization of background factors and related variables, including gender, gender identity, work experience, career stage, and leadership stereotype and how they influence achievement motivation in various contexts.

Background Factors

People's experiences in the workplace differ in many ways due to many personal characteristics, some of which are changeable, while others are less amenable to change. The background factors that are considered as primary dimensions of diversity are essentially unchangeable and exert significant impacts. They include race, ethnicity, age, sexual orientation, and physical abilities/disabilities. The primary dimensions of diversity shape our basic self-image, sense of identity and early learning experiences. These dimensions fall in to different categories, whether they are regarded as not easily changeable or unchangeable depending on society's belief and situations. Researchers often ignore the interdependence of them with other (secondary) dimensions of diversity, in which the latter are changeable personal characteristics that are acquired, and may be modified or abandoned throughout life. Although the secondary dimensions of diversity are more controllable, they will be affected by other people's decisions (Powell & Graves, 2003), which in this book we see as environmental factors. This is consistent with Farmer's (1997) conceptualization of the influences of background factors that are mostly indirect and mediated by the personal and interactive environmental factors.

Further, one of the key constructs of Bandura's (2006) theory, called symbolizing capability, could be considered as relating to the pre-influence of background factors on achievement motivation. Through the creation of symbols, such as images (mental pictures) or words, humans are capable of giving meaning, form, and contiguity to their experiences. In addition, the ability to form symbols enables individuals to accumulate information in their memory that can be used to guide future actions. It takes place through one of humankind's daily abilities, a process called "model observed behaviour" in which symbols provide the mechanism for thought (Bandura, 2001). One could argue that the primary dimension of diversity, which shapes our sense of identity, is close with mental pictures or words, gives meaning to the experience, and guides future action through shaping our personal and environmental perception. Thus, through this conceptual lens, the group of background factors is considered as affecting personal and environmental factors.

Betz and Hackett (1981), based on Bandura's social learning theory, attempted to address a self-efficacy approach to women's career development. They postulate that women's low expectations of success are a primary source of internal barriers and are influenced by the sex differences in efficacy expectations. Thus, women show less expectations of personal efficacy regarding many career-related behaviours, and so fail to fully realize their capabilities and talents in their career goals (Leal, 2006).

According to Ismail (2006), gender qualities are shaped through the various social processes, such as social relations and interactions. When the relations involve men and women, they are called gender relations, which are essentially the relations of power, dominance and hierarchy, characterizing the life structure of men and women. Therefore, gender relations are socially and culturally constituted, and they take very different forms under different circumstances and socio-cultural contexts. Thus, the first characters that shaped gender quality were gender identity and gender role stereotypes, which in the workplace were followed by occupation stereotypes.

Different background variables were found to influence the achievement motivation dimensions. For example, in Farmer's (1985; 1997) model, which worked on students, the background variables of achievement motivation included sex, social status, school location, race, age, and mathematics and verbal ability. In the present case of extension agents' leadership role, gender and gender identity variables are found to be influential. Further, the career stage, work experience, and a new variable of leadership stereotypes as a primary source of internal barriers, counted due to its consequent relationships with stereotypical assumption of gender role stereotype and its significant effect on achievement motivation on leadership role. The following sections discuss each of the background variables, including gender, gender identity, work experience, career stages, and leadership stereotypes.

Gender

Sex is a biological characteristic, while gender refers to the social role associated with being male or female. Gender is a set of socio-cultural constructs, such as roles, responsibilities, and expectations. Gender is a pervasive concept and it becomes very dynamic when it interacts with careers (Ismail, 2008).

Powell and Graves (2003) reviewed psychological literature and concluded that there are sex differences, such as those in cognitive abilities and children's interests and activities. Sex differences help males in some types of cognitive ability, such as visual-spatial abilities. Females also have advantages in most types of verbal abilities from birth to death. The extent to which sex differences differ over a life-span may have decreased over time (Powell & Graves, 2003). One could argue that female advantages in verbal ability become an important component of leadership competency, which could assist women in leadership roles, while negative effects of stereotypical assumption will cancel the overall effect on leadership.

Sex differences in children progress as they grow up, and they will have considerable differences in interest and activities through adolescence. These differences starting at an early age, for example, toy preferences emerge at about 2 years of age; more boys than girls have a preference for more physically aggressive rough-tumble manner of games. Sex differences in moral judgments consist of men tending to use a justice orientation, while women use a care orientation, which is concerned with maintaining relationships, being sensitive, and avoiding harm to others. There are many types of social behaviours that are affected by sex differences, such as males engaging more in physical aggression than females; males being more likely to present heroic aid, while females tend to offer nurturing or caring help; men tend to be more influential, while women are more easily influenced; and women have higher skills in nonverbal communication and behaviour than men (Powell & Graves, 2003).

Ismail (2008) explained that the common beliefs of sex differences are the origin of gender difference ideologies in the workplace in all societies. Actually, gender differences, as Powell and Graves (2003) revealed, symbolize beliefs about male and female roles in the community and workplace. Also gender differences are manifested in stereotypes, prejudice, and discrimination, which are mostly applied to women. Thus, it could be argued that gender does not cause differences, while stereotypical assumptions of gender differences exclude women from some occupations and positions. Thus, influencing variables on personal factor mostly were related to gender role stereotype, not gender, which will be discussed in the gender identity section.

Karami et al.'s (2009) study of Iranian post graduate students in a Malaysian university found gender did not significantly influence respondents' aspiration to achieve top management. Lopes (2006), in studying the differences in dimensions of career motivation between international and domestic graduate students at Pennsylvania State University, supported these findings of no significant relationship between gender and achievement motivation. Similarly, Akiyoshi's (2007) investigation on the effects of tertiary education on Japanese women's employment status and career aspirations showed gender did not relate to career aspiration. Contrary to the study of Cannon (2004), Farmer (1985) and London and Noe (1997) showed gender was significantly related to all three measures of career identity (London scale, $r = -.15$; Noe et al. scale, $r = -.16$; Carson & Bedeian scale, $r = -.13$, $p < .01$ for all correlations), showing that men scored higher in career identity than women (London & Noe, 1997). A possible explanation for the inconsistent results about the effect of gender may relate to recent gender similarities in educational attainment, and consequently aspiring and achieving higher positions in managerial hierarchy and leadership role.

There is further indication that there is no significant gender difference regarding leadership, motivation, and academic achievement, as Borgen and Betz (2008) found from a study on students enrolled in psychology courses at a large Midwestern university. They found that women scored significantly higher than men only on 'teamwork' ($p < .05$) with a small to moderate size effect of .37. In post-hoc analyses, men scored higher than women only on 'risk taking' ($p < .001$), with a large effect size of .83. The results showing no gender differences are similar for a study in Malaysian public sector organizations among managers (Mohd Rasdi & Ismail, 2011).

Gender was chosen as one of the background variables for several reasons: First, there are conflicting results in the literature about the influence of gender on achievement motivation. Second, the results of Karami et al.'s (2009) study were another incentive to include gender in this research. Karami et al. (2009) found that Iranian women had the same level of aspiration as men towards top managerial position, but women believed that society accepts them more in middle and lower levels of managerial positions, although they have high educational attainment. However, men have higher desire to reach top management positions than women, and feel more comfortable, as society accepts them at the top as well as in middle managerial positions.

Gender Identity (Gender Role Stereotypes)

Gender identity refers to the characteristics an individual develops or learns, rather than characteristics that appear from biological origins (O'Neill and O'Reilly, 2004). Sex-role orientation is a process by which individuals acquire sex-appropriate preferences, behaviours, and personality characteristics. Gender roles and gender identity are taught and learnt within a complex set of relationships in the family and are reinforced through the socialization process in every society (Chugh & Sahgal, 2007).

For a while masculinity and femininity were the core dimensions of human personality; each was considered appropriate and mental health is required when a person confirms to his or her gender role stereotype. It is believed that males possess highly masculine traits such as independence and dominance. Feminine traits such as gentleness and sensitivity to the feelings of others are mostly possessed by females (Powell & Graves, 2003).

Later, some researchers challenged these assumptions and beliefs. Bem, as a pioneer researcher in this area, classified gender identity based on a four-quadrant classification scheme (androgynous, masculine, feminine, and un-differentiated) and clarified that standards of mental health should be genderless, in 1974. Bem introduces androgyny, which refers to high femininity and masculinity, a more flexible response to situation and other positive factors, which were seen to better meet complex demands of society. Debates over androgyny continue in four conceptions: 1- conformity with gender stereotype, 2- break from gender stereotype and pursuit of an androgynous ideal, 3- no standards of ideal and people should be themselves, and 4- as Bem argued, behaviour should have no gender, and standards of behaviour are dependent on the culture. While these debates are continuing, gender identities have been changing, as in the U.S. where it seems to be closer to androgynous ideal (Powell & Graves, 2003).

Bandura (1989) stated that gender role is adopted through a vast system of socialization practices, beginning at birth with infants clothed in pink or blue apparel depending on their sex. The traditional gender role orientations are not uniformly sex-linked, not all males are aggressive, nor are all females unassertive, but people adopt the prevalent social examples they observe. Stereotypic gender roles that people adopt have lifelong effects on how they perceive and process experiences and how they use their capabilities.

A recent study conducted on Iranian graduate students at UPM as a potential future population of the workplace showed that gender identity explained a large proportion of career aspiration variance. The results revealed that femininity influences people's career aspirations, bringing a lower level of career aspiration; feminine women especially have lower career aspirations than feminine men. Also, the results of logistic regression analysis determined that masculinity had a significant positive effect on aspiration to top management (Karami et al., 2009), which is similar to the results of the studies by Powell and Butterfield (2003) and Hensley (2002).

O'Neill and O'Reilly (2004) suggest the importance of gender identity rather than sex as a determinant of career success or achievement. They argued that what defines an individual is the combination of sex and gender identity, because there are conflicting findings depending on whether a researcher investigates male and female differences through only sex, or gender identity (O'Neill and O'Reilly, 2004).

Thus, gender identity was found to be influential on achievement motivation of leadership role through personal and environmental factors and refers to the sex-appropriate preferences or characteristics an individual develops or learns, rather than characteristics that appear from biological origins, including masculinity and femininity traits.

Work Experience

Personal behaviour is determined when an individual acts on his or her environment, experiences the consequences of the behaviour (both negative and positive), then reflects on the situation and cognitively stores the information, which then guides future behaviour (Mills, Pajares, & Herron, 2007; Ottino, 2009). Through this process of reflective self-examination, individuals estimate the sufficiency of their thoughts, behaviour, and motivation, and modify their thinking and later their behaviour (Mills et al., 2007). Thus, experiencing a leadership role could shape subsequent behaviour and motivation in the form of achievement motivation in leadership role.

The non-occupational positions occupied before beginning the individual career stage affect the future positions (Super, 1980). One could argue that the earlier occupational positions that are occupied also influence the way in which person role expectations are met in the near future. The occupational position could also be considered as an indicator of past and current socio-economic status, which was counted as an important factor affecting personal achievement motivation by Farmer (1985). Powell et al. (2002) also studied gender and managerial stereotypes in a longitudinal manner and recommended that different managerial levels be included in future research.

Bandura's theory of vicarious learning refers to learning from direct experience or observation, which could alter efficacy beliefs through transmission of competencies and comparison with the attainment of others (Bandura, 2000). Thus, the effect of work experience on cognitive ability through life span, and consequently achievement motivation, is supported, similar to the study results of Dorsen, Carlson and Goodyear (2006) on work experiences and career aspiration of students. Therefore, work experience refers to the number of years an extension agent has worked, which could influence his or her achievement motivation in a leadership role.

Career Stage (Age)

It is predicted that in the year 2010 almost half of the global employees were forty-five years or older in most developed countries due to the aging of the baby boomers, relative decrease in birth rates, and economic situations that encourage late retirement (Kanfer, 2004). Particularly in developing countries such as Iran this is further exacerbated by the low rate of employment. The older workers exert indirect and varied impacts on job outcomes, such as their different incentives to achieve job roles such as in leadership.

A longitudinal study by London and Noe (1993), with an age-based distinction, categorized workers as trial-career workers (30 years old or younger), mid-career workers (31-45 years old), and late-career workers (46 years old and older). This study, as the first study of London and Noe, demonstrated that older workers had higher self-reported levels of career resilience. Further, in their next study, London and Noe (1997) found that personality characteristics related with achievement motivation altered at the same time as an individual ages, because of altering career experiences and adaptation to the environment. Thus, people adapt their vocational preferences and competencies with the changing situations in life and work (Super, 1980). Furthermore, Bandura's theory of social cognitive discussed the effect of experience on cognitive ability across life span (Mills et al., 2007).

Super (1980) further demonstrates that life span career stages include: first, growth (4-13 years) which means fantasy, interests, capacities; second, exploration (14-24 years) which means crystallizing, specifying, and implementing an occupational choice; third, establishment (25-44 years) meaning stabilizing, consolidating, and advancing in a position; forth, maintenance (45-65 years) which means holding on, keeping up, and innovating; and fifth, decline or disengagement (over 65 years) meaning deceleration, retirement planning, and retirement living (Super, 1980). Furthermore, individuals in different life stages have special approaches to emphasize different roles, and these approaches would affect their achievement motivation (Super, 1980; Super & Nevill, 1984). Thus, as the subjects are engaging in different career stages, their interest to achieve leadership roles could also be different.

Ottino (2009) showed that the impact of age on motivation is as varied as claimed by other researchers who have closely examined this area of influence. Some researchers believe that aspiration and mastery motivation are established at a very young age and others believe that the impact of life experience continually alters and changes them. Also, Domenico and Jones (2006) stated aspirations are often crystallized during the early stages of adolescence and the studying period (Domenico & Jones, 2006). It could be implied that an early crystallization of aspiration may cause differences in achievement motivation of individuals in different career stages.

Kanfer (2004) revealed that motivation in midlife and later years follows the same basic principles as work motivation in young adulthood, namely the allocation of personal resources to work behaviours that were built on competencies, promote a sense of self-efficacy and self-concept, and offer opportunities for the attainment of desired outcomes. The author revealed that different work environments provide different opportunities for protecting self-concept and different incentives that affect the age-related changes of employees. Hence, the differences in work motivation as adults age are mainly the impact of experiencing different age-related changes that may have increased or decreased results on work motivation.

A piece of research in an Iranian extension work context showed respondents' age had a statistically positive and significant relationship with motivation level (Sadighi, 2006). Similarly, surveys such as that conducted by London (1993) on workers with a mean average age of 57.8 years, found that some elements of career motivation increased with age. He measured career resilience and found it increasing with age ($r = .15$, $p < .05$). He also found a relationship between age and insight, whereby older workers had higher levels of career insight. But there was no significant relationship between career identity and age. The three mentioned terms, resilience, identity and insight, according to London and Noe (1997) are considered as parallel with Farmer's (1997) dimensions of achievement motivation.

The study of Greller (2000) discussed those who believed mentoring associated with career motivation and new skill development are appropriate concerns for older people who had greater career motivation. A recent piece of research also revealed that there were significant differences in achievement motivation dimensions between age groups while accounting for gender and student nationality. Women showed an increase in career resilience and identity levels throughout their entire career. For men, there was an increase in career resilience, career insight, and career identity from trial to mid-career. In addition, older women had higher levels of both insight and identity than trial-career men and women (Lopes, 2006).

Thus, career stage, which could be assessed based on respondents' ages, is selected as a background variable, because, based on Super's theory, people at establishment level have higher achievement motivation due to attempts for career advancement, as discussed earlier.

Leadership Stereotypes

A stereotype is a set of beliefs about the personal attributes of a group of people (Powell, Butterfield, & Parent, 2002). According to Bandura (1989) many of the shared misconceptions about occupational pursuits are at least partly cultivated through symbolic modelling of stereotypes. Occupation stereotypes have important effects on work self-efficacy expectations more for females than males with equivalent abilities. This idea is supported by Bandura, that occupation stereotypes would affect the personal attributes in general and self-efficacy in particular, especially for females, which consequently affect their achievement motivation (Leal, 2006). Management and leadership in the old definition are masculine tasks and their attributes are related to male dominant stereotypes. It is a stereotypical belief about managers that may be expressed as "think manager - think male", in which, as noted above, a stereotype is a set of beliefs about the personal attributes of a group of people (Powell, Butterfield, & Parent, 2002).

Currently, being a good manager is generally related to good human relation skills and mostly to quite feminine characteristics rather than masculine ones. Thus, recent organizational changes are concerned more with self-management, empowerment, and organizational learning, which call for feminine characteristics. Although in reality being more feminine does not mean less masculine. Forces that include self-selection, organizational selection, and organizational socialization all contribute to confirm leader stereotypes and reinforce a belief in the good manager as masculine (Powell et al., 2002). Similarly, a longitudinal study of Powell, Butterfield, and Parent (2002) revealed that, despite the extensive increase in the percentage of women in leadership roles over the past three decades and more emphasis on feminine traits in leadership, it is still the case that both men and women with different educational levels, ages, and work experience, consider a good manager to be a person who mainly possesses masculine traits. However, studies that directly measure leader effectiveness reject the idea that men are better leaders and that better leaders are masculine.

Leader stereotypes put women who are interested in leadership roles at a distinct disadvantage. Women who are interested in leadership roles have to hide or forget their socially desirable gender role. If women aspiring to leadership roles possess the female gender role, then they fail to meet the requirements of the leader stereotype. Further, women who do not see themselves as fitting the stereotype of a good leader may not develop leadership skills and may be diverted from pursuing a leadership role and vice versa. Thus, leadership stereotypes may lead to discrimination against women in leadership roles; they receive lower performance ratings than men and fewer promotions into higher leadership positions, because they continue to face prejudice in the leader's role, such as people feeling uncomfortable working for a female leader (Powell & Graves, 2003).

Women are ranked lower in stereotypically male dominated jobs. In this case, in the work setting, women are seen as less achievement-oriented than men. There is also confirmation that women are viewed in a different way from men when it comes to leading others. However, the majority of evidence suggests that this is because management or leadership is stereotypically a male dominated job (Connerley & Mecham, 2008).

Hoyt (2007) asserted that there are far more similarities than differences in the leadership behaviours of women and men, and they are equally effective. However, women are less effective in a masculinized leadership position such as the military, and are considered more effective in education, social services, and anywhere that communal interpersonal skills are highly valued.

Thus, leadership stereotypes are defined as a set of beliefs about the personal attributes of leadership roles. Promoting effective leadership, organizations should take advantage of individuals and give leaders of both sexes equal chance to succeed. There is no need to teach male leaders to behave more like women or vice versa. This is further supported by situational leadership theory asserting that leaders should be masculine, androgynous, feminine, and finally undifferentiated in turn as maturity of followers increases (Powell & Graves, 2003).

Conclusion

Background factors were conceptualized using the concept of primary dimensions of diversity, which are not easily changeable or unchangeable, and are not isolated from other dimensions that play an important role in a persons' achievement motivation. Having in mind the prevalence of stereotypical thinking about leadership, the background factor variables that shape gender quality, such as gender role stereotypes, which in the workplace are followed by leadership stereotypes, as well as career stage and work experience, are all considered. The importance of gender quality as the base of recognizing the variables of background factors is related to the low gender ratio of women extension workers, and to the importance of gender homophily of extension workers with clients due to cultural norms in which female extension clients should be approached by female extension workers.

References

Akiyoshi, M. (2007). Women are women are women?: The effects of tertiary education on Japanese women's employment status and career aspirations, *The Annual Meeting of American Sociological Association.*

Bandura, A. (1989). Social cognitive theory. *Annals of child development, 6*, 1-60.

—. (2000). Cultivate self-efficacy for personal and organizational effectiveness. In E. A. Locke (Ed.), *Handbook of principles of organization behavior*. Oxford, UK: Blackwell.

—. (2001). Social cognitive theory of mass communication. *Media Psychology, 3*(3), 265-299.

—. (2006). Guide for constructing a self-efficacy scales. In F. Pajares, & Urdan, T. C. (Ed.), *Self-efficacy beliefs of adolescents* United State of America: IAP-Information Age Publishing, Inc.

Borgen, F. H., & Betz, N. E. (2008). Career self-efficacy and personality: linking career confidence and the healthy personality. *Journal of Career Assessment, 16*(1), 22–43.

Cannon, S. C. (2004). *Supporting first-year college women in math, science, and related majors: A Career Development Intervention (Doctoral dissertation).* Doctor of philosophy, North Carolina State University, United States.

Chugh, S., & Sahgal, P. (2007). Why do few women advance to leadership positions? *Global Business Review, 8*(2), 351-365.

Connerley, M. L., & Mecham, R. L. (2008). Gender differences in leadership competences, expatriate readiness, and performance. *Gender in Management: An International Journal, 23*(5), 300-316.

Domenico, M. D., & Jones, H. K. (2006). Career aspirations of women in the 20th century. *Journal of Career and Technical Education, 22*(2), 18-25.

Dorsen, J., Carlson, B., & Goodyear, L. (2006). Connecting informal STEM experiences to career choices: identifying the pathway. A literature review produced by the ITEST Learning Resource Center. Retrieved 29, Jan, 2009, from http:// www.itestlrc.edc.org/connecting-informal-stem-experiences-career

Farmer, H. S. (1985). Model of career and achievement motivation for women and men. *Journal of Counseling Psychology, 32*(3), 363-390.

—. (1997). Women's motivation related to mastery, career salience, and career aspiration: A multivariate model focusing on the effects of sex role socialization. *Journal of Career Assessment, 5*(2), 355-381.

Greller, M. M. (2000). Age norms and career motivation. *The International Journal of Aging and Human Development 50*(3), 215 - 226

Hensley, L. (2002). *The Influence of Gender Role and Gender Stereotyping on the Career Choice and Career Commitment of Adolescents*. Retrieved 29 Jan, 2009 http://www.samford.edu/schools/artsci/scs/hensley.html.

Hoyt, C. L. (2007). Women and leadership. In P. G. Northouse (Ed.), *Leadership theory and practice (fourth ed.)*. Thousand Oaks, London, New Delhi: SAGE Publication.

Jivetti, B. A., & Edwards, M. C. (2009). *Selected factors affecting the performance of women's self-help group in Western Kenya.* Paper presented at the 25th Annual Conference of AIAEE "Twenty-five Years of Strengthening International Agricultural and Extension Education", May 24 to 27, InterContinental San Juan Resort, Puerto Rico.

Kanfer, R. (2004). Aging, adult development, and work motivation. *Academy of Management Review, 29* (3), 440-458.

Karami, R., Ismail, M., & Md.Sail, R. (2009). *Gender identity on career aspiration to top management among Iranian graduate students of an institution of higher learning.* Paper presented at the 1st International Conference on Educational Research and Practice, Malaysia.

Leal, V. M. (2006). *Predictors of non-traditional career self-efficacy in Mexican American adolescent women (doctoral dissertation).* PhD, The Ohio State University USA.

London, M. (1993). Career motivation of full- and part-time workers in mid and late career. *The International Journal of Career Management, 5*, 21-29.

London, M., & Noe, R. A. (1997). London's career motivation theory: An update on measurement and research. *Journal of Career Assessment, 5*(1), 61-80.

Lopes, T. (2006). *Differences in dimensions of career motivation between international and domestic graduate students by age and gender (doctoral dissertation).* Doctor of Philosophy The Pennsylvania State University United States.

Ismail, M. (2006). Conflicting understanding about gender and development In J. Othman, R. M. Sail & A. Ahmad (Eds.), *Women and peace; issues, challenges and strategies*. Selangor: Institute for Community and Peace Studies and Women's Studies Unit, Faculty of Educational Studies, University Putra Malaysia.

—. (2008). *Gender & career: realities and challenges.* Inaugural lecture series at UPM, Serdang: University Putra Malaysia Press.

Mohd Rasdi, R., & Ismail, M. (2011). *Managerial success in the Malaysian public sector organizations: Does gender matter?* Paper presented at the 12th International Conference, Sustaining Growth Through Human Resource Development, 25-27 May, The University of Gloucestershire Cheltenham, England.

Mills, N., Pajares, F., & Herron, C. (2007). Self-efficacy of college intermediate French students: relation to achievement and motivation. *Language Learning 57*(3), 417–442.

Navarro, M., & Ricketts, J. C. (2008). *Strong village, strong country: leadership development of future farmers of Georgia adolescence participating in a week-long leadership training institute* Paper presented at the 24nd Annual Conference of AIAEE "Global Entrepreneurship in International Agricultural and Extension Education", Costa Rica, pp. 376-384.

O'Neill, A. O., & O'Reilly, A. C. (2004). *Women's careers: the impact of sex and gender identity on career attainment.* Research paper No. 1775(R). Graduate School of Business; Stanford University.

Ottino, K. L. (2009). *Diminished aspiration: women central office administrators and the superintendency (doctoral dissertation).* PhD, The University of Minnesota United State.

Powell, G. N., & Butterfield, D. A. (2003). Gender, gender identity, and aspirations to top management. *Women in Management Review, 18*(1/2), 88-96.

Powell, G. N., Butterfield, D. A., & Parent, J. D. (2002). Gender and managerial stereotypes: have the times changed? *Journal of Management, 28* (2), 177-193.

Powell, G. N., & Graves, L. M. (2003). *Women and men in management.* Thousand Oaks, London, New Delhi: SAGE Publications.

Sadighi, H. (2003). *Extension re-organizational engineering commensurate with progresses in technology* Paper presented at the 19 Annual Conference of AIAEE, Raleigh, North Carolina, USA, pp. 560-569.

Sadighi, H. (2006). *Factors affecting agricultural extension personnel's motivation level.* Paper presented at the 22nd Annual Conference of AIAEE "International Teamwork in Agricultural and Extension Education", May 14 to 19 Clearwater beach, Florida.

Super, D. E. (1980). A life-span, life-space approach to career development. *Journal of Vocational Behavior, 16*, 282-298.

Super, D. E., & Nevill, D. D. (1984). Work role salience as a determinant of career maturity in high school students *Journal of Vocational Behavior 25*, 30-44

CHAPTER FOUR

PERSONAL FACTORS AND ACHIEVEMENT MOTIVATION IN THE LEADERSHIP ROLE

Personal factors refer to the perception regarding a person's cognitive processes, such as belief about ability that acts as an important mediator of achievement motivation, and influences expectations for future performance. The personal factor is conceptualized using the concept of secondary dimensions of diversity and related variables in the context of extension for achievement motivation in leadership role. This chapter discusses the influence of self-efficacy, self-esteem, achievement value, and leadership competencies on the achievement motivation in leadership role.

Personal Factors

Personal factors are related to individuals' beliefs, abilities, and emotions (Alderman, 2004), or in the words of Bandura in social cognitive theory, personal factors are cognitive and biological events, as well as affective domain (Bandura, 1999). Cognition is self-perceptions, beliefs, and expectations that shape and direct individual behaviour. Biological events are related to physical structure and sensory and neural systems (brain structures) that affect people's behaviour. The affective natural and extrinsic effects of actions also partly determine individual thought patterns and emotional reactions (Bandura, 1989). In Bandura's perspective, motivation involves the interrelation of the three factors, including cognition-personal, environment, and the behaviour or performance of the person. These three factors interact through a process that Bandura called reciprocal interaction, in which each component affects the other two, even though they may not be of equal strength. In this view, the person's cognitive processes, such as belief about his or her ability, act as important mediators of motivation, influencing both expectations for future performance and action taken (Alderman, 2004). Personal factors are in fact referred to as secondary dimensions of diversity which are acquired and may be modified throughout life, and are more controllable by an individual than the primary dimension of diversity (Powell & Graves, 2003).

Farmer (1985), by using Bandura's theory, concludes that the personal set, more than the environment set, contributed toward salience and mastery dimensions of achievement motivation, although in the aspiration dimension of achievement motivation, background, environment, and personal order were more suitable. Farmer (1985; 1997), based on her longitudinal study, found different personal variables affecting achievement motivation. Some variables that were included in Farmer's studies were dropped in the present discussion because they did not obtain strong significant relations with the achievement motivation dimensions. The other reasons for not discussing some of the variables under the personal factor are, for example, Farmer's model of achievement motivation focuses on high school students to investigate the factors predicting career achievement motivation, while our focus in this analysis is to explain the influencing factors on achievement motivation of leadership role among extension agents. These differences emerge when, for example, Farmer (1985) counts the effects of self-concepts as expressive and independent, which are developed during adolescence, whereby men first develop a sense of independence and then expressiveness, while women have the reverse procedure. Thus, for people at the working age, perhaps both senses developed earlier. The other dropped variables are achievement style, success-failure attribution, and personal unconcern.

There are two different styles of achievement, consisting of competitive and cooperative. The competitive situations increase mastery motivation for men and vice versa for women. Competitive achievement style is a significant contributor for both sexes in aspiration, while cooperative achievement style significantly affects salience. However, in basic psychological research about gender differences of career achievement, achievement styles are very interesting; they are not considered in this discussion.

Success-failure attribution is internal in origin and comes from people's beliefs relating achievements to ability or effort, rather than to external sources such as luck or other people (Farmer, 1985). Success-failure attribution seems important for an investigation of academic achievement because of its special effect on self-esteem (Bandura, 2006; Daniels & Edwards, 2004; Farmer, 1985). Further, it could affect students' motivation for job achievement through its indirect effect from self-esteem to mastery motivation (Daniels & Edwards, 2004). But the extension agents achieved their career and position as extension agents, and they struggle in achieving an emergent job with a leadership role. Also, Farmer

(1985; 1997) counted this variable because at that time it was not yet well developed in the literature. But nowadays such a psychological concept of external or internal locus of control that a person ascribes for his or her success-failure obviously affects achievement motivation and is not counted in the present discussion. The same explanation is true for the variables such as personal unconcern. In fact, personal unconcern refers to individuals' concern for the points of view of others, which is mostly meaningful in achieving an assigned position, not an emergent leadership role.

Based on the reviewed literature, self-esteem, self-efficacy, achievement values, and leadership competency, are considered as the variables of personal factors affecting achievement motivation in the leadership role of extension agents, in which Farmer's (1985,1997) and Bandura's theories (1989, 1999, 2000, 2001, 2006) support the importance of all mentioned variables.

Self-Efficacy

Self-efficacy in the history of psychology, acknowledged as an important development (Borgen & Betz, 2008; Pajares & Urdan, 2006), originated from Bandura's Social Learning Theory. Self-efficacy is defined as people's beliefs regarding their ability to effectively carry out a given task (Bandura, 2006). Bandura asserted that self-efficacy is the major mediator between people's behaviour and behavioural change (Hampton, 2005). So it is believed that phenomena such as human motivation, learning, self-regulation and accomplishment (Pajares & Urdan, 2006), and well-being and success in life (Burke, 2005), could not be explained without discussing the function of self-efficacy. Bandura further explains that different contexts in which people develop their efficacy shape different levels of self-efficacy, which are unique to an individual (Bandura, 2006). Self-efficacy beliefs are cultivated through four information sources, namely vicarious learning, emotional arousal, performance accomplishments, and verbal persuasion (Leal, 2006).

Bandura asserted that self-efficacy is a major cognitive mechanism in which an individual commits themselves to a valuable goal, looks for self-satisfaction from achieving the goal, and then tries to make a greater effort (Bandura, 2001). Individuals with higher self-efficacy beliefs concentrate on the opportunities that are valuable to them, whereas individuals with low self-efficacy beliefs think that they have little control on environmental barriers and their efforts are ineffective, and consequently have lesser effort and achievement (Bandura, 2000).

The literature about self-efficacy shows that there is extensive support for the role of self-efficacy in determining work related performance (Steers et al., 2004). Cox (2008) found that self-efficacy is essential for women to aspire to move into leadership positions and likewise to feel confident in their abilities for performing this role. Karami, Ismail, Omar, and Wahat (2011) also found self-efficacy is positively related to achievement motivation.

Findings of a recent study showed the effect of self-efficacy on work-related performance, control personality, intelligence or general mental ability, and job or task experience. Although a meta-analysis of the relevant literature in this area of research explained that self-efficacy mostly predicts job or job role performance with low complexity, it could not predict those jobs or job roles with medium or high complexity. There is research literature discussing the predicting role of self-efficacy on task performance as part of a job, but not overall job performance. In general, individual differences and environmental situations could affect the predictive validity of self-efficacy (Judge, Jackson, Shaw, Scott, & Rich, 2007). This predictive effect of self-efficacy for task performance, rather than overall job performance, brings the present analysis to concentrate on one job role; that is, leadership.

Larson et al. (2007) studied Taiwanese students and found that female students had less self-efficacy in traditionally male-dominated majors and occupations compared with male students. Further, they found that self-efficacy significantly contributed to students choosing majors and their future choice goals, such as career aspirations. Leal (2006) also found that aspirations and mastery motivation were positively related to self-efficacy in non-traditional careers. In addition, she verified a significant moderation effect for ethnic identity in the relationship of mastery motivation and non-traditional career self-efficacy.

Ramly, Ismail, and Uli's (2009) study found a positive linear relationship between self-efficacy, organizational socialization and continuous improvement practices, and the professionals' career aspirations among R & D professionals. They further found self-efficacy and continuous improvement practices as the main contributors in explaining verification in career aspiration.

According to Bandura (2006), there are diverse measures for self-efficacy. The Career Decision Self-Efficacy Scale (CDSES) established by Taylor and Betz in 1983 is one of the frequent measures. It consists of 50 items (tasks) and covers five domains of career choice competence. These five domains include goal setting, problem solving, self-appraisal, occupational information, and planning. The Career Decision Self-

Efficacy Scale was shortened by Betz, Klein, and Taylor in 1996 with the same domains in 25 items (Hampton, 2005). Also the General Self-Efficacy Scale (GSE) was developed for the adult population and intended to appraise a general sense of perceived self-efficacy. The German version of GSE was developed by Matthias, Jerusalem and Ralf Schwarzer in 1979, which has also been adapted to 26 other languages by different researchers. Thus, GSE has been utilized in different studies and reported a Cronbach's alpha ranging from .75 to .91 with the majority at more than .80 (Burke, 2005).

Self-efficacy will be used in this analysis referring to individuals' (extension agents') beliefs about their capability to successfully perform a given task, that is, leadership role.

Self-Esteem

According to Bandura (2006), self-esteem is a judgment of self-worth. Zimmerman and Cleary (2006) are in agreement with Bandura in which self-esteem is a kind of engaging assessment of self-worth and how individuals feel about themselves. Self-esteem could be built up from a person's general self-perception or from a range of different sources in society. The reciprocal determinism of Bandura's personal and environmental factors are supported in the words of Zimmerman and Cleary (2006), stating that a person who perceives a sense of competence may establish feelings of self-worth and consequently is affected by acceptance of society. The opposite is also true, that the self-esteem of a person affects the perceived environmental situation.

Farmer (1985; 1997) stated that individuals with high estimates to do academic tasks are more likely to have higher career achievement motivation. Further, she found a positive and significant relationship between academic self-esteem and salience dimension of career achievement motivation.

People's beliefs about themselves have important consequences, for example, Ladebo, Olaoye, and Adamu (2008) revealed that higher self-esteem, which is people's belief about their self-worth, could affect job satisfaction and commitment. They reported the detail of this relationship as follows: organizational commitment ($\beta = 0.58$, $p < 0.001$), supervisor commitment ($\beta = 0.28$, $p < 0.001$), and workgroup commitment ($\beta = 0.33$, $p < 0.001$). They conclude that people with high self-esteem have higher commitment to the extension organization, extension supervisor and workgroup (Ladebo et al., 2008) but simultaneously bring higher commitment to the job and job roles. They suggested that in recruitment of new personnel, extension managers have to consider their self-esteem and make sure that their personnel expand their self-esteem through:

> - "showing sympathy to extension personnel, when they have personal problems or crises and also, to value their opinions and contributions to work;
> - assisting extension agents to form realistic views about their abilities and encouraging them to always accept the responsibility for their decisions;
> - providing work assignments that offer challenges, autonomy, and variety to match extension personnel's skills and competences;
> - building trust among co-workers and between managements and extension personnel; and
> - providing clearly defined work expectations and feedbacks on performances" (Ladebo et al., 2008, p. 261).

Self-esteem is affected by the acceptance of society and thus affects individuals' perceived environmental situation. The results from studies of Josephs, Bosson, and Jacobs (2003) showed people with low self-esteem are less interested in accepting positive feedback from themselves than from an external source, but they are interested in accepting negative feedback from both self and external resources. One could argue that organizations have to be aware of positive feedback from an external source, which is highly accepted by high and low self-esteem individuals. Further, employees with low self-esteem accept negative feedback from both internal and external sources, while they only accept the positive feedback from external sources, leading them to have more powerful sources for accepting considerable negative feedback. Thus, according to Josephs, Bosson, and Jacobs (2003), positive feedback from outsiders may temporarily increase the self-esteem of individuals who doubt their own self-worth and later might even result in permanent increases of self-esteem. Furthermore, the effect of an outsider's feedback on individual self-esteem and vice versa is an example that shows the reciprocal relationships of self-esteem and environmental factors, as proposed by Bandura. Hein's and Hagger's (2007) study results showed that global motivational orientations, such as intrinsic motivation, affect self-esteem and later affect autonomous motivation. In sum, self-esteem could be defined operationally as an individual judgment of self-worth to handle a leadership role.

Achievement Values

According to Oberhauser et al. (2004) reward, as a term connected to reinforcement, is anything an individual perceives as valuable. Intrinsic and extrinsic are two universal types of rewards. Intrinsic rewards are approachable by feelings of achievement, challenge, and self-esteem which are related to the work itself. The sense of having an important effect, a friendship, or social interaction are also considered as intrinsic rewards. Extrinsic rewards are achieved through funding of the organization (Oberhauser et al., 2004). Maehr and Nicholls, cited in Farmer (1985), identified three types of values underlying achievement motivation, including: intrinsic, competence, and social approval. Intrinsic values "involve doing something for its own sake, whereas competence values imply doing something for utilitarian reasons such as to obtain a skill". Social approval values "involve doing something for external approval or recognition" (Farmer, 1985, p. 368).

Farmer (1985), in her previous research through investigating intrinsic, competence, and social approval, only found a significant relationship for the intrinsic value. Thus, only intrinsic value is considered as achievement value in Farmer's studies. Similar to most of the studies that measured success and achievement, Farmer went through objective variables. But it is considerable that female and male use different types of evaluation value in assessing career achievement. Men were most likely to prefer objective measures rather than subjective measures, while women are the reverse (Powell and Graves, 2003). One could argue that, although Farmer (1997; 1985) was interested in understanding gender differences in achievement motivation, she did not consider the predominant achievement value of women as subjective value.

Objective value refers to receiving more pay, higher position and advancing at a faster rate. Subjective value refers to a person's satisfaction with different aspects of work and non-work life, such as job security, family friendly policies, and opportunities to peruse hobbies and other personal interests (Powell & Graves, 2003). People are not only different at the level of motivation, but also different in orientation of the motivation. Orientation of motivation concerns the underlying attitudes and goals that give rise to action (Ryan & Deci, 2000), such as achievement. Based on the different reasons or goals of motivation Deci & Ryan in 1985 developed Self-Determination Theory (SDT) in which there are different types of motivation orientation or values, classified as intrinsic and extrinsic motivation. They also discussed the motivation orientation as unwillingness for a behaviour (Ryan & Deci, 2000). Thus, we consistently consider intrinsic and extrinsic motivation as values that objectively motivate individuals, but it is further categorized that unmotivated individuals are, in part, persons with subjective achievement value.

Super (1980) hypothesized that playing different roles at the same time would result in role conflict, commitment to one role making it difficult to do justice to another. One could argue that this is the reason why some individuals, especially women, have subjective achievement value and men objective value. According to Powell and Graves (2003) this comes back to their predominant gender role and traditional responsibilities of them that women are home makers and men are breadwinners. Women generally place more attention than men on their own feelings and on their relationships with others, and less emphases on whether they are getting ahead (Powell & Graves, 2003), which guides them to the subjective value. The effect of subjective achievement value in the present discussion is further supported by Farmer's (1985) explanation regarding negative effects of competing activities stemming from family and home interests on the salience dimension of achievement motivation.

Value theory postulates 10 types of basic values that were categorized based on two main goals as: how an individual expresses personal goals, and how an individual interacts with others to pursue collective goals (Schwartz & Rubel, 2005). The study results of Dio, Saragovi, Koestner, and Aube (1996) using value theory also showed that men are more likely to endorse agentic values, such as freedom and accomplishment, whereas women are more likely to endorse communal values, such as friendship and equality. Further, in another study, they indicated masculine values were found to be significantly related to socially desirable masculine traits, similar to the feminine values that were found to be significantly related to socially desirable feminine traits. Thus, it is interpreted that values expressing personal goal are considered as objective and the values expressing a collective goal are subjective.

Objective value includes intrinsic and extrinsic values. Intrinsic motivation is considered as a vital value, showing the natural human trend to learn, understand and achieve (Ryan & Deci, 2000). Through assessment of the extent to which individuals' intrinsic and extrinsic motivations are strong and significant to them, and the extent to which individuals vary in those motivations, we could better understand and predict motivational behaviour (Amabile, Hill, Hennessey, & Tighe, 1994). Jong (1999) discussed intrinsic motivation in terms of engagement in work. Amabile, Hill, Hennessey, and Tighe (1994) revealed intrinsic motivation value connected to desire for personal enrichment, challenge, enjoyment, and self-determination. There are employees who look at their jobs as opportunities to activate their competencies and skills, follow personal interests and make meaningful contributions to society (Vansteenkiste et al., 2007), which here we consider as employees with intrinsic achievement value.

Extrinsic motivation is argued to be a sign of external control or true self-regulation (Ryan & Deci, 2000). Jong (1999), defined extrinsic motivation as the motivation to work to attain something distinct from the work itself, such as reward or recognition. Vansteenkiste, Neyrinck, Niemiec, Soenens, Witte, and Broeck (2007) found that having extrinsic compared to intrinsic job value orientation was associated with a smaller amount of positive job outcomes and extra negative job outcomes. An extrinsically motivated employee's focal point is financial success, controlling and influencing others, and occupying a prestigious position (Vansteenkiste et al., 2007); here we consider this group of employees as persons with extrinsic achievement value.

Amabile, Hill, Hennessey, and Tighe (1994) stated that individuals' intrinsic and extrinsic motivation were theoretically discussed as moderately stable individual differences, but could change as a function of environmental situations. They suggested that extrinsically motivated individuals tend to show more impatience, rigid behaviour in task engagement, poorer concept attainment, impaired complex problem solving, poorer incidental learning, increased functional fixedness on a set-breaking task, and lower levels of creativity in a variety of tasks. This group more likely looks to the organizational support for extrinsic motivation. The authors also mentioned that intrinsically motivated individuals are interested in choosing work tasks that permit new skills development, practice of creativity, and deep involvement in the work. This group is more likely to look to the work environmental situation support for their intrinsic motivation, and they might look for occupations where intrinsic motivators are prominent.

A variety of previous studies has investigated the relationship between job value orientations and positive job outcomes, such as job commitment. Nevertheless, limited studies have investigated the relationship between extrinsic as compared to intrinsic job value orientations and negative job outcomes (Vansteenkiste et al., 2007). Further, the assessment of achievement values, consisting of subjective and objective (intrinsic and extrinsic), would be a novelty in this area of research. The achievement value in this book is operationally defined as an individual's subjective and objective value underlying achievement of leadership role. The subjective achievement value refers to importance and priority given to personal interests such as work-family balance in achieving leadership role. Objective achievement value refers to persistence in importance of extrinsic or intrinsic achievement value. Intrinsic value refers to the importance of achieving a leadership role as an opportunity to exercise competencies and skills, pursue personal interests and make meaningful contributions to society. Extrinsic value refers to the importance of achieving a leadership role to attain something distinct from the work itself, such as financial success, having control and influence over others, or occupying a prestigious position at work.

Extension Agents' Leadership Competencies

The philosophy and principles of extension are used as the basis for extension workers to function and hence in developing their roles and consequently their professional competencies (Quinn & Boone, 1964). Basically, the role of extension agents is to improve farm practices, and activities in the home as well as the community. Where the social, economic, and political situations have evolved, the roles of extension agents have also evolved from someone who merely provided the content of technological information to a facilitator who smooths the running program, and creates a vision (Morse et al., 2006); functions that are parallel with leadership role of extension agents, who also require such competencies.

Extension agents have multiple functions and the degree of emphasis on each function varies depending on situation. Each agent acts as an organizer and planner, a teacher, an interpreter of research findings, an educational consultant, an educational catalyst, an educational spokesman, a liaison person, a developer of leaders, an educational statesman, and as an evaluator (Quinn & Boone, 1964). Actually it is not possible to describe the role of the extension agents in a few words. They are expected to carry out the following functions (Allo, 2007, p. 2):

- "Help farmers improve their living standards;
- Help farmers achieve their long and short term objectives;
- Make practical suggestions which will enable farmers to attain their goals;
- Act as a link between farmers, researchers and planners;
- Help farmers to devise methods of overcoming their problems. This is an innovative function, and is also problem-oriented;
- Assist in the implementation of national policies;
- Assist in the organization of farming structures."

The above mentioned functions of extension agents to the farmers are all about a teaching and learning process expected to empower the clients. To empower the clients, extension agents need to be empowered first, in terms of knowledge, attitude, skill, and practice. Then, empowered extension agents can build the capacity of clients and the latter can adopt good farm management and farm technology; being, in one word, empowered. Empowerment is a milestone in extension work. Empowerment of agricultural extension agents requires them to be trained on the visible (technical and management knowledge of an enterprise) as well as the hidden (motive, traits, self-concept) aspects of their job competency, one of which is leadership (Md.Sail, & Asnarulkhadi, 2010).

Md.Sail (2010) defines the role of extension agents as to provide extension services to clients in the form of technical and management knowledge, as well as providing human resource development knowledge and practices of capacity building to empower clients to solve their own problems and make their own decisions. Md.Sail (2005) further discusses the role of extension agents is to transfer technology and develop the clients capacity and potential (human resource development) to enable them to make decision and solve their problems, and finally to play leadership roles to develop their own community.

Allo (2007) discusses the role of extension agents beyond technological expertise and notes the need for a new role to deeply understand people. This means an extension worker's role goes beyond technical innovation motivator, to a local leader to help in mobilizing resources in the society (Khalil et al., 2008). Thus, in order to effectively empower a client, extension staff should receive training in leadership development before transferring knowledge and skills to rural communities (Arnold & Meyers, 2007). The acceptance of a leadership role by extension personnel assists in assuring sustainability of rural community development, through which leadership is experienced in activities such as rural cooperatives (Lawrence, 2000).

Quinn and Boone (1964) explained that extension workers must provide dynamic educational leadership for clients. This means helping people to make an inventory of their physical and human resources, analyse problems, identify objectives, consider alternative courses of action, and develop programs aimed at solving their problems. It is obvious that developing the leadership potential of the people is the responsibility of an extension worker; hence they need to have effective leadership competencies.

Scheer, Ferrari, Earnest, and Connors (2006) presented a model for preparing extension professionals. They illustrated the model in which the competencies required for success in extension profession are as follows:

- Extension knowledge, leadership, and management
- Technology
- Communications
- Program planning, implementation, and evaluation
- Applied research
- Diversity and pluralism
- Marketing and public relations
- Theories of human development and learning
- Risk management
- Community development process and diffusion

Thus, leadership competency is considered as one of the core competencies of extension professionals (Scheer et al., 2006). Burke (2002) stressed seven leadership competencies for extension workers, namely problem solving, coaching, organizational skills, team goals, personal-interaction skills, verbal communication, written communication.

Ladewig and Rohs (2000) defined leadership competencies of extension agents, including managing the job, building the team, thinking clearly, and relating to others. Managing the job could include time management, setting goals, planning, and scheduling work. Building the team is defined through training, coaching, and delegating, appraising people and performances, disciplining and counselling. Thinking clearly was interpreted as identifying and solving problems, making decisions, and weighing risk. Relating to others includes skills in listening, giving clear information, and getting unbiased information (Ladewig & Rohs, 2000).

Khalil (2008) reported that more than half of the extension agents (57.2%) demonstrate a moderate level of perceived leadership competency, whereas 22.1% and 20.7% indicated a high and low level, respectively. He hypothesized leadership as one of the required extension agent competencies and a part of extension work, and found that extension competencies had a positive and significant relationship with extension worker performance ($r=.621$; $p<.006$).

Bandura's social cognitive theory supports the importance of knowledge, abilities and skill which are determinant of competency on people's behaviour. Bandura argued on acquisition of knowledge and skills as requirements for cognitive attainment. Thus, knowledge and skills or, overall competency, play an influential role in cognitive development

(Bandura, 1989) and consequently cause people to behave in a certain way. The way people behave based on their competencies, for example in attainment of a leadership role, affects their environment in a reciprocal determinism, which guides understanding of causal relationships and expanding knowledge through symbols derived from personal and vicarious experiences (Bandura, 2001).

Overall, leadership competency is considered as one of the most important competencies of extension agents and it is operationally described as the knowledge, abilities and skills that are used by extension workers to guide the clients.

Conclusion

Agricultural extension climate in developing and developed countries is influenced by the quality of the extension staff. While previous studies have integrated different variables within personal factors that motivate a person to achieve, this chapter provides a substantiation of the personal factors affecting achievement motivation in leadership roles of the agricultural extension agents.

The identified variables of personal factors, in addition to applicability for future research, could assist extension organizations to develop appropriate training programs. The training program on leadership competencies, self-efficacy and achievement values are important due to their direct and indirect effect on achievement motivation, as the need for achievement could be taught. Further, the identified variables in the personal factor have implications for organizations in staff selection and recruitment. The self-efficacy, achievement value, and leadership competencies could become criteria to be used in the recruitment process. Therefore, extension organizations require or expect their extension agents to get engaged in effective leadership roles.

References

Alderman, M. K. (2004). *Motivation for achievement: possibilities for teaching and learning*. New Jersey: Lawrence Erlbaum Associates, Inc.

Allo, A. V. (2007). The training of extension workers. Retrieved 2 November, 2009, from Food and Fertilizer Technology Center publication database www.www.agnet.org/library/eb/200a

Amabile, T. M., Hill, K. G., Hennessey, B. A., & Tighe, E. M. (1994). The work preference inventory: assessing intrinsic and extrinsic motivational orientations. *Journal of Personality and Social Psychology, 66*(5), 950-967.

Arnold, S., & Meyers, C. (2007). *Leadership styles of international agricultural extension agencies.* Paper presented at the 23rd Annual Conference of AIAEE "Internationalizing With Cultural Leadership", May 20 to 24, Polson, Montana.

Bandura, A. (1989). Social cognitive theory. *Annals of child development, 6*, 1-60.

—. (1999). Social cognitive theory: An agentic perspective. *Asian Journal of Social Psychology 2*, 21-41.

—. (2000). Cultivate self-efficacy for personal and organizational effectiveness. In E. A. Locke (Ed.), *Handbook of principles of organization behavior*. Oxford, UK: Blackwell.

—. (2001). Social cognitive theory of mass communication. *Media Psychology, 3*(3), 265-299.

—. (2006). Guide for constructing a self-efficacy scales. In F. Pajares, & Urdan, T. C. (Ed.), *Self-efficacy beliefs of adolescents* United State of America: IAP-Information Age Publishing, Inc.

Borgen, F. H., & Betz, N. E. (2008). Career self-efficacy and personality: linking career confidence and the healthy personality. *Journal of Career Assessment, 16*(1), 22–43.

Burke, M. U. (2005). *Family influence on self-efficacy: exploring the relationship between perceived parenting style, current social support, and self-efficacy belief in a sample of nontraditional college students (doctoral dissertation).* Ph.D., Capella University, Minnesota.

Burke, T. B. (2002). *Defining competency and reviewing factors that may impact knowledge, perceived importance and use of competencies in the 4-H professional's job (doctoral dissertation).* Ph.D., North Carolina University Raleigh.

Cox, K. S. (2008). *Motivational factors influencing women's decision to pursue upper-level administrative position in higher education (doctoral dissertation).* Ph.D., Kansas State University Manhattan, Kansas

Daniels, H., & Edwards, A. (2004). *The routledgefalmer reader in psychology of education.* London: Taylor and Francis group.

Dio, L. D., Saragovi, C., Koestner, R., & Aube, J. (1996). Linking personal values to gender. *Sex Roles, 34*(9-10), 621-636.

Farmer, H. S. (1985). Model of career and achievement motivation for women and men. *Journal of Counseling Psychology, 32*(3), 363-390.

—. (1997). Women's motivation related to mastery, career salience, and career aspiration: A multivariate model focusing on the effects of sex role socialization. *Journal of Career Assessment, 5*(2), 355-381.

Hampton, N. Z. (2005). Testing for the structure of the career decision self-efficacy scale-short form among Chinese college students. *Journal of Career Assessment, 13* (1), 98–113.

Hein, V., & Hagger, M. S. (2007). Global self-esteem, goal achievement orientations, and self-determined behavioural regulations in a physical education setting. *Journal of Sports Sciences, 25*(2), 149 – 159.

Jong, E. d. (1999). The impact of motivation on the career commitment of Dutch literary translators from Elsevier Science www.elsevier.nl/locate/poetic

Josephs, R. A., Bosson, J. K., & Jacobs, C. G. (2003). Self-esteem maintenance processes: Why low self-esteem may be resistant to change. *Personality and Social Psychology Bulletin, 29*(7), 920-933.

Judge, T., Jackson, C., Shaw, J., Scott, B., & Rich, B. (2007). Self-efficacy and work-related performance: the integral role of individual differences. *The Journal of Applied Psychology 92*(1), 107-127.

Karami, R., Ismail, M., Omar, Z., & Wahat, N.-W. A. (2011, 25-27 May). *Influence of mentoring on achievement motivation in leadership role as perceived by extension workers in Iran.* Paper presented at The 12th International Conference, Sustaining Growth Through Human Resource Development The University of Gloucestershire Cheltenham, England.

Khalil, A. H. O. (2008). *Relationships between extension competencies, organizational commitment and job satisfaction with performance of agricultural extension workers in Yemen (doctoral dissertation).* Doctor of philosophy, University Putra Malaysia, Selangor.

Khalil, A. H. O., Ismail, M., Suandi, T., & Silong, A. D., (2008). Extension worker as a leader to farmers: influence of extension leadership competencies and organizational commitment on extension workers' performance in Yemen. *The Journal of International Social Research, 1*(4), 368-387.

Ladebo, O. J., Olaoye, O. J., & Adamu, C. O. (2008). Extension personnel's self-esteem and workplace relationships: Implications for job satisfaction and affective organizational commitment foci. *The Journal of Agricultural Education and Extension, 14*(3), 249-263.

Ladewig, H., & Rohs, F. R. (2000). Southern extension leadership development: Leadership development for a learning organization. *Journal of Extension, 38*(3).

Larson, M. L., Wei, M., Wu, T. F., Borgen, H. F., & Bailey, C. D. (2007). Discriminating among educational majors and career aspirations in Taiwanese undergraduates: the contribution of personality and self-efficacy. *Journal of Counseling Psychology, 54*(4), 395-408.

Lawrence, L. D. (2000). *A 21th century challenge: extension role in economic development of rural communities.* Paper presented at the 16nd Annual Conference of AIAEE "Partnerships with the Private Sector", March 29 to April 1, Arlington, Virginia.

Leal, V. M. (2006). *Predictors of non-traditional career self-efficacy in Mexican American adolescent women (doctoral dissertation).* PhD, The Ohio State University, USA.

Morse, R. S., Brown, P. W., & Warning, J. E. (2006). Catalytic leadership: reconsidering the nature of extension's leadership role. *Journal of Extension, 44*(2).

Oberhauser, C. L., Egea, M., Gilbert, S., & Lopez, A. (2004). Motivation theory Retrieved 29 May, 2009, from *userwww.sfsu.edu/~gtarakji/engr801/.../motivation.html*

Pajares, F., & Urdan, T. C. (2006). *Self-efficacy beliefs of adolescents.* United States of America: IAP-Infornation Age Publishing, Inc.

Powell, G. N., & Graves, L. M. (2003). *Women and men in management.* Thousand Oaks, London, New Delhi: SAGE Publications.

Quinn, E. H., & Boone, E. J. (1964). Raising our professional sights in extension. *Adult Education Quarterly, 15*(23), 23-34.

Md.Sail, R. (2005). *Towards an effective extension delivery system for rubber smallholders in Malaysia*. Paper presented at the International Rubber Research and Development Board TOT Workshop on Smallholders, April 4 to 8, Kuala Lumpur, Malaysia.

—. (2010, 26-28 October). *Empowering of agricultural extensionists through knowledge and skill of the extension profession.* Paper presented at the International Conference on Agricultural Extension, AGREX 10: Empowerment of Agri-Food Stakeholders in Facing Global Challenges Towards Sustainability Palm garden IOI hotel & resort, Malaysia.

Md.Sail, R., & Asnarulkhadi, A.-S. (2010). Community development through community capacity building: A social science perspective. *Journal of American Science, 6*(2), 68-76.

Ramly, E. S., Ismail, M., & Uli, J. (2009). Antecedents of career aspiration of R&D professionals in Malaysian Public Organizations. *European Journal of Scientific Research, 26*(1), 66-79.

Ryan, R. M., & Deci, E. L. (2000). Intrinsic and extrinsic motivations: Classic definitions and new directions. *Contemporary Educational Psychology, 25*, 54-67.

Scheer, S. D., Ferrari, T. M., Earnest, G. W., & Connors, J. J. (2006). Preparing extension professionals: The Ohio State University's model of extension education. *Journal of Extension, 44*(4).

Schwartz, H. S., & Rubel, T. (2005). Sex differences in value priorities: Cross-cultural and multimethod studies. *Journal of Personality and Social Psychology 89*(6), 1010-1028.

Steers, R. M., Mowday, R. T., & Shapiro, D. L. (2004). Introduction to special topic forum the future of work motivation theory. *Academy of Management Review, 29*(3), 379-387.

Super, D. E. (1980). A life-span, life-space approach to career development. *Journal of Vocational Behavior, 16*, 282-298.

Vansteenkiste, M., Neyrinck, B., Niemiec, C. P., Soenens, B., Witte, H. D., & Broeck, A. V. d. (2007). On the relations among work value orientations, psychological need satisfaction and job outcomes: A self-determination theory approach. *Journal of Occupational and Organizational Psychology, 80*, 251-277.

Zimmerman, B. J., & Cleary, T. J. (2006). Adolescents' development of personal agency: The role of self-efficacy beliefs and self-regulatory skill. In F. Pajares & T. C. Urdan (Eds.), *Self-efficacy beliefs of adolescents.* United States of America: IAP- Information Age Publishing, Inc.

CHAPTER FIVE

ENVIRONMENTAL FACTORS AND ACHIEVEMENT MOTIVATION IN THE LEADERSHIP ROLE

Individual learning takes place in environmental situations through observational learning, imitation, and modelling, which are important variables in explaining the environmental factors of achievement motivation. This chapter covers the supportive environmental variables, including organizational supports, supervisors' supports, and mentoring, which is a determinant in modelling of human behaviour. However, there are many different environmental variables affecting achievement motivation, such as political, cultural, and social variables. Also, in this chapter, by considering men's and women's desire, the supportive environment that leads to achievement of excellence in work-related behaviour is considered as a situation in which both work and family life are supported.

Environmental Factors

The social cognitive theory from Bandura supports the effect of environmental situation on motivation and what motivates an individual to behave in a special way, such as attainment of excellence or achievement of leadership as one of the key roles in extension work. According to Bandura's social cognitive theory, there are three types of environmental structure, involving imposed, selected, and constructed environment; people's control over each environmental situation increases respectively. But it is important that the potential environmental situation that is actually experienced is based on people's personal agencies and the way people behave (Bandura, 1999). Bandura discusses the interaction of social context as environmental variables and cognitive variables that shape individuals behaviour (Amarasing, 2008; Bandura, 2001). Thus, theoretically the environmental situations, such as support or deviance, are important factors in achievement motivation. In this chapter, with a positive view, just the supportive environment variables are considered, including organization and supervisor supports. In addition to the mentioned variables, and considering the focus of Bandura's theory on the process of human behaviour through modelling, the other variable called mentoring was shaped in the investigation of environmental factors.

Farmer's model (1997) recognizes three environmental influences related to achievement motivation, all of which engage support systems that include parent and teacher support in addition to support for women working. She showed the positive effect of family and parental support on children's achievement motivation. Also she found that support for women working is the most important variable to predict career aspiration as a dimension of achievement motivation.

The other main theorizers in the area of workers' motivation were London and Noe (1997) who demonstrated the importance of situational conditions on career motivation. They found that motivation components were higher in an organizational environment that supported career development (London & Noe, 1997). One question that needs to be asked, however, is whether only work support is enough to motivate a person for achievement of work related behaviour. In our understanding, by considering men and women in the mind, a supportive environment that leads to achievement of excellence in work related behaviour is a situation in which both work and family supports exist together.

In Powell and Graves' (2003) view, there are three general principles that an organizational situation should follow to maximize employees' opportunities to have successful careers and higher achievement motivation for both men and women. They include bias free decision, access to developmental experiences, and a family friendly environment. To consider the three mentioned principals in this study, work and non-work support, along with meritocracy, were counted as the dimensions of organization and supervisor supports. While meritocracy is representative of bias free decision, work and non-work supports are representatives of access to developmental experiences and a family friendly environment, respectively.

Thus, this chapter concentrates on the supportiveness of extension organization climate as the foundation of environmental factors. The organizational climates will be discussed through extension organization supports and extension organization supervisors' supports, which both include work and non-work dimensions. Mentoring, which is one of the advised methods in leadership empowerment, was also added as a variable that could be introduced as an alternative manner of workplace learning.

Mentoring

The origins of the term mentoring are in Greek mythology, and have mostly been related to the teaching and learning process, which has an experienced and older person acting as the role model of others in a medium of learning (Mincemoyer & Thomson, 1998; Strouse, Sieverdes, & Hecht, 2005). In the workplace, a mentor is defined as an experienced employee who acts as a role model and gives feedback, support, and guidance for the mentee and who is committed to the mentee's success. Two comprehensive mentoring functions of mentors include career and psychosocial functions. Career functions are related to sponsorship, coaching, exposure/visibility, protection, and the provision of challenging assignments. Psychosocial functions include role modelling, counselling, friendship, and acceptance (Day & Allen, 2004). According to Noe (1988) psychosocial functions of mentoring affect job involvement and career planning activity. While there are different studies that researched the role of mentoring, looking at the mediation effects of mentoring in the processes involving achievement motivation of leadership roles would be a novelty.

The vicarious capability construct of Bandura's theory supported the importance of mentoring in achievement motivation. Vicarious processes speak about the human ability of learning from direct experience and observation of others. Observational learning enables a person to build up an initiative opinion of behaviour formation, without actually practicing a behaviour. Then, this information and opinion is expressed in symbols and is implied as a lead for future action called vicarious capabilities (Bandura, 2000). Observational learning is useful for discovering conditions that commonly are out of reach to be achieved, as a result of time, resources, and mobility limitations. The observer is mostly selective and goes to the modelling of people's behaviours of people who are most like themselves (Stone, 1998). A mentor as a role model could simultaneously transmit new ways of thinking and behaving to many people in different areas (Bandura, 1989) as raw data, which in cognitive processing becomes experience, as theorized by Bandura (2000). Thus, mentoring as an environmental factor variable, in addition to directing the effect on a mentee's behaviour in a reciprocal relationship with cognitive abilities of the mentee, could affect behaviour. One such behaviour is achievement motivation (Bandura, 1989). Kram (1983) asserted that an experienced mentor through psychosocial function will be a good source for raising a mentee's self-esteem, self-confidence, and sense of competence.

The literature supports the importance of mentoring in career advancement in general and in leadership development in particular. A leadership mentor has the function of a unique landmark in an individual's life (Strouse et al., 2005). According to Ottino (2009) the current broad research on leadership concurs that mentoring is vital to the evolvement of leadership abilities and fundamental to career advancement. Although in the case of women aspiring to leadership, lack of a mentor emerged as a barrier. One could argue that lack of a female mentor is the consequence of stereotypical thinking about leadership roles.

There are different aspects that have to be taken into account in mentoring. In a mentor and mentee relationship there are some important aspects that have to be considered, such as gender preferences and culture difference, in addition to the time spent in mentoring relationship and applicability of formal or informal relationships (Strouse et al., 2005). Strouse, Sieverdes and Hecht (2005) discussed the effect of gender roles and age on the mentee-mentor relationship and beneficial results of a formal mentoring program for the mentee, the mentor, and the organization.

The mentoring source varies at different life stages, for example family friends, coaches, religious leaders, business operators, teachers, and community leaders could be mentoring sources in teenage and young adult years (Strouse et al., 2005). But questions that need to be asked are: Who is the source of mentoring in the workplace; supervisors or pairs? And in which situation; formal or informal? London and Noe (1997) revealed that working for a boss that serves as an ideal role model and supports employees' development may be a source of resilience, insight, and identity development. It is a true and important finding that a good boss could be a role model and act as a source of motivation for achievement, but still the effect of similarity or dissimilarity of position in organizational hierarchy between mentor and mentee is questionable.

The research results about similarity of mentor and mentee gender are inconsistent, for example, Strouse, Sieverdes and Hecht (2005) found that a male mentee largely self-selected male mentors, while female mentees equally choose male and female mentors. One recent study concluded that opposite sex mentoring has advantages and also hazards. The advantages of opposite sex mentoring for women is an "in", or in other words, an entrance to the good old boy network. While hazards for women in opposite sex mentoring may happen when the good old boys have the power to damage women's dignity, intelligence, stature, and credibility (Ottino, 2009). Overall, in the case of women, mentoring has a significant effect on self-efficacy and enhances their chance of career progression. Thus, women need to be aware of mentoring importance and need to make a mindful attempt to create support groups with colleagues or friends in the organization or outside of their organizations (Cox, 2008).

Ricciardi (2005) investigated the outcome of a one-year mentoring program conducted for entry-level administrators. They analysed mentoring outcome through the Noe (1988) mentor function scale and found that the participants who experienced more hours in mentoring relationships rated higher career and psychosocial support. However, respondents' gender, race, position and school level, and mentor and mentee similarity did not significantly affect the mentoring outcomes. The author further suggested some consideration for policy makers in designing mentoring programs that should consist of supporting both functions of mentoring, policies regarding duration, mentor training and pairing, and program goals (Ricciardi, 2005).

Researchers argued that mentor relationships take place in a series of successive events in a lifetime, each with its own life cycle and time-line (Kram, 1983). Strouse, Sieverdes and Hecht (2005) identified four stages in the protégé-mentor relationship. The first stage is related to the beginning of the mentee-mentor relationship. The second stage is a stage of debate between two parties and the finding of a new manner of acting and attitudes. Acquisition of a new manner of acting and attitudes by a mentee is the starting point of the third stage, in which most of the significant learning takes place. During this stage the protégé develops confidence and gains a sense of empowerment for moving. In stage four, the mentee will redefine the self and look for more independence. However, the question that needs to be asked is whether each stage is related to a certain age or duration of age, or are these stages repeated in each of the mentoring relationships which were discussed here.

Mincemoyer and Thomson (1998) studied a planned mentoring program in Pennsylvania State Cooperative Extensions. They found that when both mentor and mentee are accountable in the same programs, it provided more favourable time for communication during program planning and more chance for interaction and professional progression activities. Geographic nearness of mentors and protégés was recognized as a facilitative feature. Successful mentoring relationships were found to be related to a shared variety and frequency of information, particularly about program development or technical/subject information. Also, the initiation of contact through face-to-face mentoring meetings at the first available time and in a friendly manner seems to assist the victory of the mentoring relationships. But in this modern age, geographical factor is no longer an important matter, since there is possibility for e-mentoring to happen. Simmonds & Lupi (2010), in investigation of an innovative e-mentoring program, support the influence of mentoring on achievement of mentee learning for whom time and geographical limitations may prevent face to face meetings and traditional off-job training.

Thus, mentor refers to an experienced employee who serves as a role model with career functions, including sponsorship, coaching, exposure/visibility, protection and the provision of challenging assignments; and psychological functions, including role modelling, counselling, friendship and acceptance for a low experienced employee.

Extension Organization Supports

Perceived organizational support is defined as perception of employees regarding the organizational concern for their well-being (Chow, Lo, Sha, & Hong, 2006). Organizational support theory revealed that workers develop attitudes regarding the extent to which the organization and its agents (e.g., leaders) think and care about their happiness and welfare (Eisenbergera, Huntingtona, Hutchisonb, & Sowa, 1985; Ferris, Brown, & Heller, 2009). It is perceived that organizational support influences an employee's emotional connection and as a result their commitment to the organization, which is consistent with social exchange theory. To the extent that perceived organizational support increases, work effort of workers for meeting organizational goals will also be increased, which may be related to raising mastery as a dimensions of achievement motivation (Eisenbergera et al., 1985).

Extension organizations, similar to any other organization, need to have extension agents and workers satisfied with their jobs. Consequently, they will show higher connections and commitment to the organizations, the mangers, and co-workers (Ladebo et al., 2008). High levels of perceived organizational support generate a sense of responsibility to pay back the helpfulness of organization, which will be shown through extra effort and role behaviour (Eisenbergera et al., 1985; Hochwarter, Kacmar, Perrewe, & Johnsonc, 2003). One could argue that organizational support may affect mastery motivation through increased effort in job roles.

People are more motivated to achieve a job role when they find relevant supportive policies in the organizations for their achievement efforts. This is similar to Xu's (2007) work about more probability of participation in developmental activities, such as taking courses when people find relevant supportive policies in the organizations for their development efforts. Thus, organizations have to support opportunities of different job roles, performance and acquisition of confidence for better usage of abilities, and finally provide developmental opportunities to the employees (Chow et al., 2006).

According to London and Noe (1997), dimensions of achievement motivation are affected by perceived organizational supports. Organizational support characteristics that influence career resilience include positive reinforcement and constructive performance feedback, encouragement of autonomy, opportunities for individual control and discretion. Organizational support characteristics which influence career insight consist of goal setting structure, path goal structure, organizational flexibility, opportunity for change, and visibility of organizational processes. Situational characteristics that influence career identity include encouragement of professionalism, pressing for organizational commitment, advancement opportunities, potential for recognition, leadership opportunities, and potential for monetary gain (London & Noe, 1997).

Chow, Lo, Sha, and Hong (2006) found that developmental experience had significant effect only on performance, while organizational support intensifies workers' sense of pride, which will lead to raising results. Thus, one could conclude that organizational support through affecting the employee's sense of pride would motivate them to achieve. Perceived organizational support is a determinant of work outcome and personal commitment, such as in the study of Hochwarter et al., (2003) which used a general scale of organizational support. Work environment and the importance of its support are influential in satisfaction study as well. In such a study, McCusker, Dendukuri, Cardinal, Katofsky, and Riccardi (2005) identified four aspects of work environment, namely supervisory support, team-work, professionalism, and interdisciplinary relations. Xu (2007) considers organizational work support, non-work support, supervisors supports, and co-workers supports as contextual variables for supportive environment. Although the meaning of non-work support in Xu (2007) is supportive role of family and friends, here the supportive role of organization in family friendly policies was used instead.

A family friendly environment could be defined as a work environment that leans toward having work and family balance. Promoting a family friendly environment needs an organizational culture in which taking time out from career for family reasons does not reflect a lack of career commitment (Powell &Graves, 2003). The family friendly environment supported in the study of Ogden, McTavish, & McKean (2006), includes child care, parental leave, and flexible working. Thus, family-supportive environments refer to a situation in which flexitime, part-time work, and on-site day-care are available (Lapierre et al., 2008).

The level of work life balance is very important to an individual with subjective achievement value, which mostly applies to women or feminine characters. The findings of Ogden, McTavish and McKean (2006) supported the idea that work-life balance is a considerable factor in women's structural barrier for achievement of a leadership role. They said that a lot of females are interested in a logical work-life balance, but, due to the structural barrier, believe that they are not a homogenous group in comparison with men for achievement of leadership roles.

The study of Lapierre, Spector, Allen, Poelmans, Cooper, and O'Driscoll (2008) discussed the effect of family-supportive environment on the decreased level of work-family conflict. They also revealed that the lesser level of work-family conflict could be interpreted as superior job and family satisfaction, followed by superior overall life satisfaction. It is argued that higher satisfaction may motivate individuals to achieve higher level of commitment to the job and organization. This is consistent with the study of O'Neill, Harrison, Cleveland, Almeida, Stawski, and Crouter (2009), who found significant relationships between work-family supportive environment, and organizational commitment. Thus, it could be concluded that organizational commitment, as an indicator of salience dimension of achievement motivation, is affected by an organizational work-family friendly climate which supports both family and work.

Individuals with objective achievement value are suspected to need meritocracy organizational support as a part of work support. Meritocracy is defined as a characteristic of a system in which rewards and advancement (status) are based on individual merit. Although the effect of meritocracy on achievement motivation is logically obvious, it is not well developed in the literature. Thus, while employees with objective value believed that good performance should be rewarded, the organization was confident that reward adequacy would guide a perseverance of positive performance (Hochwarter et al., 2003). In contrast, less organizational support, which entitled organizational deviance, represents converse behaviours of workers to the organizational norms (Ferris et al., 2009).

The present section hypothesizes that the extent to which perceived organizational support increase would influence the higher motivation to attainment of excellence in leadership role. Originally extension organization support refers to two types of work and non-work supports. As discussed earlier, work support influences mastery achievement motivation in addition to influences on commitment, and non-work support has shown influence on commitment, which is parallel with the salience dimension of achievement motivation. Thus, extension organization support refers to extension agents' perception of the extension organization's concern about their well-being, consisting of work support and non-work support. Extension organization work support is related to the environment that supports the work goal structure and attainment along with professionalism, development opportunities, and meritocracy. Extension organization non-work support is related to a family friendly environment. This refers to a work environment that desires a balance between work and family.

Extension Organization Supervisor Supports

Employees look at organization supervisors as representations of the entire organization (Hochwarter et al., 2003) and perceive supervisors supports as important in making an appropriate environment that individuals are motivated in and committed to. The exchange theory supported the importance of employee and employer relationships on commitment. According to the exchange theory, when employers and employees consider each other's needs and interests, the sense of obligation goes further than the parameters of formal responsibility, and over periods of time reinforces the linking of employees and employers (Hochwarter et al., 2003).

Supportive managers are crucial agents in creating a supportive environment, because they are the ones who make rules and regulations for the employees. Having a family supportive supervisor reduces work-family conflict and gives employees more encouragement to stay with the organization and motivates them to achieve job roles in particular and higher performance in general (Powell & Graves, 2003).

Managers or supervisors need to persuade and give support to their staff for different reasons, such as routine and stressful nature of jobs. In order for managers to motivate employees, they should provide positive feedback and show the importance of the employees. To create a good work environment, mangers should provide opportunities for staff growth and involvement in decision-making, and rewarding experiences (Chow et al., 2006). One could conclude that managerial work and meritocracy support provide opportunities for growth since they are some of the characteristics of an appropriate environment (Chow et al., 2006). This is similar to the concept of autonomy encouragement, which London and Noe (1997) have as work support indicators. Furthermore, London and Noe (1997) revealed that encouragement of autonomy and performance feedback influence career resilience, which is parallel with the salience dimension of Farmer's model of achievement motivation.

Bandura (1999) revealed that perceived social support (one of which is supervisor supports) bi-directionally operates in human adaption and change behaviour. Further, a supportive environment could enhance personal efficacy. However individuals who perceive higher self-efficacy create a more supportive environment for themselves. The effect of supervisor supports, similar to the organizational support, would be direct and bidirectional through affecting personal factors on achievement motivation.

Supervisors could support workers through the provision of feedback and resources for skill development, and encouraging exercise of recently acquired skills (Xu, 2007). London and Noe (1997) asserted that providing feedback could affect the salience dimension of achievement motivation. However, skill development and encouraging use of newly acquired skills also have the same direction of influence on salience and mastery. O'Neill et al. (2009) supported that perceived supervisors supports are related to organizational commitment. They illustrated that employees with higher perceived supervisor supports for work and family roles, reported greater organizational commitment. Thus, supervisor supports refer to the helpful role of extension supervisors as crucial agents in creating a supportive work and non-work environment.

Conclusion

This chapter, which was on environmental factors and related variables, has explored implications for nurturing and developing extension agents in organizations. Having employees who possess relevant attributes, such as self-efficacy and leadership competencies (personal factor variables), may not be sufficient to motivate them in leadership achievement, if the environmental supports do not co-exist. The source of environmental supports, as discussed in this chapter, could be the organization or supervisors, while there are researchers who also worked with the peers, family, and friends. These supports include work supports such as work goal structure, professionalism, development opportunities, and meritocracy, and non-work supports such as initiatives to balance work and family. Thus, the need in providing the appropriate strategy and teaching process to prepare work support and family friendly culture (non-work support) in extension organizations is one of the important practical implications of this chapter.

Further, development of mentoring programs can help extension agents in career functions. Mentoring provides them with opportunities to learn new skills and understand the unwritten rules, assist in meeting the deadlines of given tasks, and suggesting specific strategies for achieving career goals. At the same time, psychological functions of mentoring as a source of modelling, counselling, friendship, and acceptance, would be beneficial to the extension agents in achieving a job role, such as a leadership role.

References

Amarasing, P. K. (2008). Social learning theory. Retrieved 8, Jan, 2009, from Online Learning Laboratory; University of South Alabama www.usaoll.org/mobile/theory.../social_learning_theory.htm

Bandura, A. (1989). Social cognitive theory. *Annals of child development, 6*, 1-60.

—. (1999). Social cognitive theory: An agentic perspective. *Asian Journal of Social Psychology 2*, 21-41.

—. (2000). Cultivate self-efficacy for personal and organizational effectiveness. In E. A. Locke (Ed.), *Handbook of principles of organization behavior*. Oxford, UK: Blackwell.

—. (2001). Social cognitive theory of mass communication. *Media Psychology, 3*(3), 265-299.

Chow, I. H.-s., Lo, T. W.-c., Sha, Z., & Hong, J. (2006). The impact of developmental experience, empowerment, and organizational support on catering service staff performance. *Hospitality Management 25*, 478-495.

Cox, K. S. (2008). *Motivational factors influencing women's decision to pursue upper-level administrative position in higher education (doctoral dissertation).* PhD, Kansas State University Manhattan, Kansas.

Day, R., & Allen, T. D. (2004). The relationship between career motivation and self-efficacy with protégé career success. *Journal of Vocational Behavior, 64*, 72-91.

Eisenbergera, R., Huntingtona, R., Hutchisonb, S., & Sowa, D. (1986). Perceived organizational support *Journal of Applied Psychology, 71*(3), 500-507.

Farmer, H. S. (1997). Women's motivation related to mastery, career salience, and career aspiration: A multivariate model focusing on the effects of sex role socialization. *Journal of Career Assessment, 5*(2), 355-381.

Ferris, D. L., Brown, D. J., & Heller, D. (2009). Organizational supports and organizational deviance: The mediating role of organization-based self-esteem. *Organizational Behavior and Human Decision Processes, 108*, 279-286.

Hochwarter, W. A., Kacmar, C., Perrewe, P. L., & Johnsonc, D. (2003). Perceived organizational support as a mediator of the relationship between politics perceptions and work outcomes. *Journal of Vocational Behavior, 63* 438-456.

Kram, E. K. (1983). Phases of the mentor relationship. *Academy of Management Journal, 26*(4), 608-625.

Ladebo, O. J., Olaoye, O. J., & Adamu, C. O. (2008). Extension personnel's self-esteem and workplace relationships: Implications for job satisfaction and affective organizational commitment foci. *The Journal of Agricultural Education and Extension, 14*(3), 249-263.

Lapierre, L. M., Spector, P. E., Allen, T. D., Poelmans, S., Cooper, C. L., O'Driscoll, M. P., Kinnunen, U. (2008). Family-supportive organization perceptions, multiple dimensions of work–family conflict, and employee satisfaction: A test of model across five samples. *Journal of Vocational Behavior, 73*, 92-106.

London, M., & Noe, R. A. (1997). London's career motivation theory: An update on measurement and research. *Journal of Career Assessment, 5*(1), 61-80.

McCusker, J., Dendukuri, N., Cardinal, L., Katofsky, L., & Riccardi, M. (2005). Assessment of the work environment of multidisciplinary hospital staff. *International Journal of Health Care, 18*(7), 543-551.

Mincemoyer, C. C., & Thomson, J. S. (1998). Establishing effective mentoring relationships for individual and organizational success *JournaL of Extension, 36*(2).

Noe, R. A. (1988). An investigation of the determinants of successful assigned mentoring relationships. *Personnel Psychology, Volume 41*(3), 457 - 479.

O'Neill, J. W., Harrison, M. M., Cleveland, J., Almeida, D., Stawski, R., & Crouter, A. C. (2009). Work–family climate, organizational commitment, and turnover: Multilevel contagion effects of leaders. *Journal of Vocational Behavior, 74*, 18-29.

Ogden, S. M., McTavish, D., & McKean, l. (2006). Clearing the way for gender balance in the management of the UK financial services industry Enablers and barriers. *Women in Management Review, 21*(1), 40-53.

Ottino, K. L. (2009). *Diminished aspiration: women central office administrators and the super intendancy (doctoral dissertation).* PhD, The University of Minnesota United States.

Powell, G. N., & Graves, L. M. (2003). *Women and men in management.* Thousand Oaks, London, New Delhi: SAGE Publications.

Ricciardi, D. (2005). *Socializing and supporting new leaders: examining the mentor and Protégé relationship* Paper presented at the Democracy in Educational Leadership: The Unfinished Journey Toward Justice, November Nashville, TN.

Simmonds, D., & Lupi, A. M. Z. (2010). The matching process in e-mentoring: a case study in luxury hotels. *Journal of European Industrial Training, 34*(4), 300-316.

Stone, D. (1998). Social Cognitive Theory. Retrieved 19, Sep, 2009, from *www.infosihat.gov.my/.../Social%20Cognitive%20Theory.doc*

Strouse, S. K., Sieverdes, C. M., & Hecht, U. S. (2005). *Leadership mentor selection, developmental stages, and personal leadership development.* Paper presented at the 21st Annual Conference "Educational, Extension and Research Strategies for a Changing World", San Antonio.

Xu, Q. (2007). *A predictive model of employee self development: the effect of individual and contextual variables (doctoral dissertation).* Phd, University of California, Irvine.

PART II

A STUDY IN IRAN: FACTORS INFLUENCING ACHIEVEMENT MOTIVATION THE IN LEADERSHIP ROLE OF EXTENSION AGENTS

CHAPTER SIX

THE STUDY: METHODS, RESPONDENTS, AND ORGANIZATION

The first chapter of the book identified one of the main gaps in extension context, that is, variation in knowledge regarding extension agents' achievement motivation in leadership role, in which little is known about this phenomenon in Iran. To cover the knowledge gap, different theoretical and practical literature was reviewed, as presented in Chapters 2 to 5, and consequently, a conceptual framework had been drawn. To validate the proposed conceptual framework, a study was conducted among Iran's extension agents. This chapter presents the methodological procedures of the study, involving research design, research framework, research objectives, hypothesis, population and sampling, sample size, instrument development, validity and reliability, data collection, data analysis, and operational definitions.

Research Design

In this study a quantitative paradigm is implemented by considering the nature of the research problem, the type and amount of knowledge already available (Williams, 2003), and desire for generalization of findings to the population (Holmes, Dahan, & Ashari, 2006). Further, from the two types of survey research design according to the time of data collection, the cross-sectional design was selected over the longitudinal design. "The cross-sectional surveys study a cross section (sample) of a population at a single point in time" (Ary, Jacobs, & Razavieh, 2002, p. 379). Malhotra (2004, p. 80) also asserted that "cross-sectional design involves the collection of information from any given sample of population elements only once." Kumar (2005), in agreement with the above mentioned definition of cross-sectional design, illustrated that cross-sectional study design is well-matched to studies expected to find out the commonness of a phenomenon, condition, problem, attitude or issue.

It has been decided to utilize a survey as the common method of social research, with three defining characteristics that make it suitable for the present research, including the type of content, the form of the data, and the method of analysis employed. The survey method is suitable for gathering data in a social context which deals with human behaviour, knowledge, attributes, beliefs and attitudes. Surveys produce a structured data set in the form of a variable-by-case grid. The form of data in the survey is systematic, structured, and based on variables, and the methods of analysis rely on comparisons across groups and relationship (Jupp, 2006). Surveys play an important role for collecting large-scale data about different aspects of social life in statistical information form. This method can be used to obtain multi-faceted data for an individual behaviour, belief, attitude and attribute. The process involves measuring various phenomena and drawing conclusions about any relationship between them, and include three main types of data collection method for social surveys: face to face interview, postal questionnaire, and telephone interview (Henn, Weinstein, & Foard, 2006). In this study, with a quantitative paradigm and implementation of cross-sectional survey design, a questionnaire has been used for data collection; part of the data is collected via a drop and picks procedure, and others through postal questionnaire.

Research Framework

Figure 6-1 depicts the variables to be examined in the study. The framework portrays the pre-influence of background factors as the independent variables, along with the mediating role of personal and environmental factors on the extension agent's achievement motivation in leadership role. The background factors consist of career stage, gender, gender identity, leader stereotypes, and work experience, which become independent variables (exogenous variables) of this research framework.

The group of personal factors consists of self-esteem, self-efficacy, achievement value, and extension agent's leadership competency. This group of variables is treated as mediator in the relationships between background factors and achievement motivation in leadership role.

The environmental factors are expected to mediate the relationship between background factor and achievement motivation in leadership role. The environmental factors include mentor supports, extension organization supports, and extension organization supervisor supports.

The dependent variable (endogenous variable) in this study is achievement motivation in leadership role, which is measured in terms of an extension agent's aspiration, mastery, and salience.

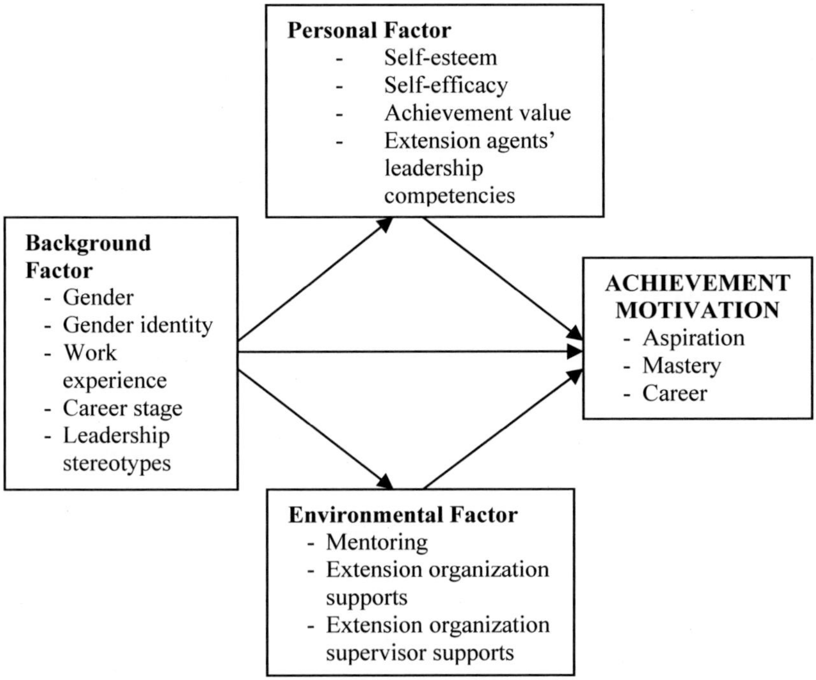

Figure 6-1: The research framework

Research Objectives

The general objective of this study is to "identify factors influencing achievement motivation in the leadership role of extension agents in Iran". The specific objectives are as follows:

1. To determine the level of achievement motivation in the leadership role of extension agents in Iran.

2. To identify the relationship between background, personal, and environmental factors and achievement motivation in the leadership role of extension agents in Iran.

3. To identify the mediation effects of personal and environmental factors on the relationships between background factor variables and achievement motivation in the leadership role of extension agents in Iran.

Research Hypothesis

In this study, based on the above research framework, four hypotheses were provided in which each hypothesis is divided into sub-hypotheses as follow:

H1: There is a significant relationship between background related variables (gender, gender identity, work experience, career stage (age), and leadership stereotypes) and achievement motivation in the leadership role of extension agents.

Hypothesis one is further divided into sub-hypothesis labelled as H1-1 to H1-5:

H1-1: There is a difference among male and female extension agents in terms of achievement motivation in leadership role.
H1-2: There is a difference in the relationship of gender identity (masculinity and femininity) and achievement motivation in the leadership role of extension agents.
H1-3: There is a significant relationship between work experience and achievement motivation in the leadership role of extension agents.
H1-4: There is a significant relationship between age and achievement motivation in the leadership role of extension agents.
H1-5: There is a significant relationship between leadership stereotypes and achievement motivation in the leadership role of extension agents.

H2: There is a significant relationship between personal related variables (self-efficacy, self-esteem, achievement value, and extension agents' leadership competencies) and achievement motivation in the leadership role of extension agents.

Hypothesis two is further divided into sub-hypothesis labelled as H2-1 to H2-4:

H2-1: There is a positive relationship between self-efficacy and achievement motivation in the leadership role of extension agents.
H2-2: There is a positive relationship between self-esteem and achievement motivation in the leadership role of extension agents.

H2-3: There is a difference in the relationship of achievement value (intrinsic and extrinsic as objective achievement value and subjective achievement value) and achievement motivation in leadership role of extension agents.
H2-4: There is a positive relationship between extension agents' leadership competencies and their achievement motivation in leadership role.

H3: There is a positive relationship between environment related variables (mentoring, extension organization supports, and extension supervisor supports) and achievement motivation in the leadership role of extension agents.

Hypothesis three is further divided into sub-hypothesis labelled as H3-1 to H3-3:

H3-1: There is a positive relationship between mentoring and achievement motivation in the leadership role of extension agents.
H3-2: There is a positive relationship between extension organization supports and achievement motivation in the leadership role of extension agents.
H3-3: There is a positive relationship between extension supervisor supports and achievement motivation in the leadership role of extension agents.

H4: Personal and environmental factors mediate the relationship between background factors and achievement motivation in the leadership role of extension agents.

Hypothesis four is further divided into sub-hypothesis labelled as H4-1 to H4-5:

H4-1: Personal and environmental factors mediate the relationship between masculinity and achievement motivation in the leadership role of extension agents.
H4-2: Personal and environmental factors mediate the relationship between femininity and achievement motivation in the leadership role of extension agents.
H4-3: Personal and environmental factors mediate the relationship between work experience and achievement motivation in the leadership role of extension agents.

H4-4: Personal and environmental factors mediate the relationship between age and achievement motivation in the leadership role of extension agents.
H4-5: Personal and environmental factors mediate the relationship between leadership stereotype and achievement motivation in the leadership role of extension agents.

Population and Sampling

The population of the study consists of extension agents of Iran Ministry of Jihade-e-Agriculture. The population of the study is scattered across a wide geographic region. Iran is segmented into thirty one provinces (Ostān). The provinces are segmented into counties (Shahrestān), and sub-segmented into districts (Bakhsh), sub-districts (Dehestān), and villages (Iran-Statistics-Center, 2009). Figure 3-2 shows a simplified organizational chart of Ministry of Jihade-e-Agriculture. This Ministry, to cover clients all over the country, has 31 Organizations of Jihad-e-Agriculture across the provinces, having a section under the title of extension office, which is the place where the study population is working. The Organization of Jihad-e-Agriculture of each province has divisions in all the counties of that province, called Administration of Jihad-e-Agriculture, with the same office for extension. In the districts, sub-districts, and villages, the extension job is done by Agricultural Extension and Services Centres, although the number of centres is not equal to the number of districts, sub-districts, and villages of each county. The area covered by each centre is determined by the population of clients and geographical accessibility to the clients (see Figure 6-2). Thus, the target population of this study are extension agents whose job is to approach clients and, according to Khalil et al. (2008), perform tasks beyond merely technical innovation motivation to human resource development leaders to help in mobilization of resources in the community.

Sampling is a method of choosing units from a population so that by studying the sample, the findings are generalized back to the population (Trochim, 2006). Since the population of this study is scattered all around a large country such as Iran, it would be better to study the subjects in naturally occurring groups or clusters, which is called cluster sampling. In cluster sampling the chosen units are mostly not individuals, but rather a group of individuals who are naturally together. These individuals constitute a cluster insofar as they are alike with respect to characteristics relevant to the variables of the study. The essential requirement of cluster sampling is to randomly choose the clusters from a population of clusters. Another requirement is that once a cluster is selected, all the members of the cluster must be included in the sample. Lastly, since the sampling error in cluster sampling is much greater than in simple random sampling, a small number of clusters makes the likelihood of sampling error greater still, even if the number of subjects is large (Ary et al., 2002). In this study, the requirements of the considered cluster sampling include: randomly selecting the clusters, considering all the members of each selected cluster as research participants, and choosing a large number of clusters.

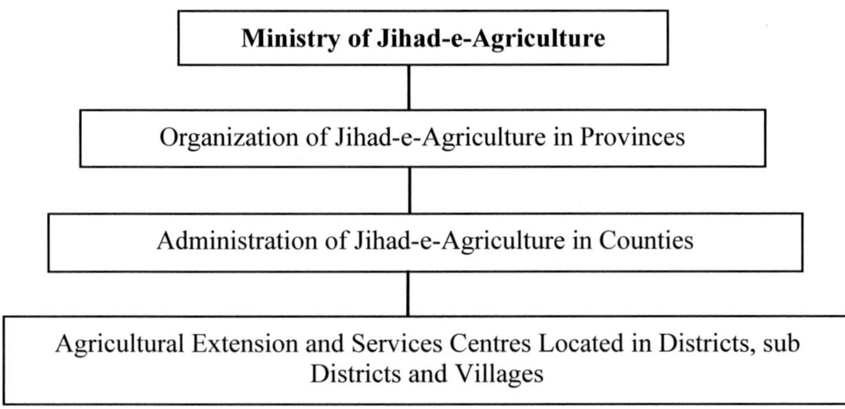

Figure 6-2: Organizational chart of ministry of Jihad-e-Agriculture

Since the study population is scattered over a broad geographical region, the study sampling was made through cluster sampling to cover a wide geographical area. The cluster sampling is structured precisely for this kind of situation. The processes of conducting cluster sampling (Figure 6-3) are as follows:

The sampling process started by randomly selecting the main clusters of study. Three Organizations of Jihad-e-Agriculture in provinces are selected randomly to participate in the study, namely Kermanshah, Tehran, and Khorasan Razavi. The number of extension organizations has been decided to meet a sufficient sample size by considering the similarity of structure and proportion of staff to clients in all of the extension organizations. Thus, approximately 10 percent of the Organization of Jihad-e-Agriculture in provinces is selected for the main clusters. Then, in each province, all Administrations of Jihad-e-Agriculture in counties and the Agricultural Extension and Services Centres located in districts, sub districts, and villages, are considered as the sub clusters of that province. Lastly, all extension agents in the selected provinces are main clusters, and sub clusters of each province are considered as the research respondents.

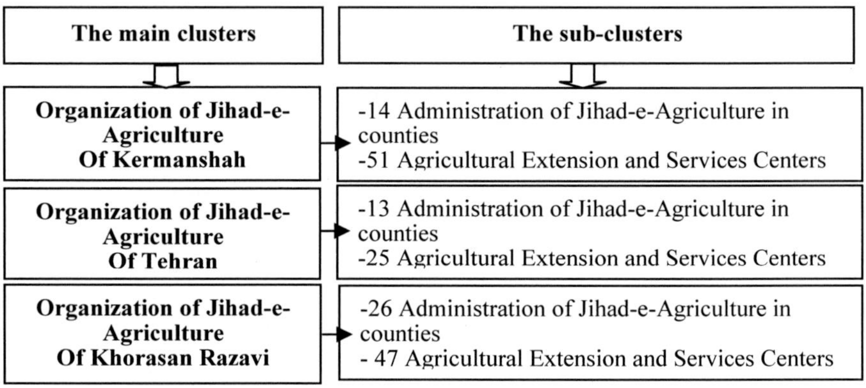

Figure 6-3: Sampling procedure of the study

Sample Size

The old ways of determining sample size were based on the size of the population. A recent method is that the test type of data analysis affects the sample size. G*Power software, as one of the recent methods, determines the sample size based on the effect size, power, and alpha error probability in a particular test. In Structural Equation Modelling there are also resources that discuss the importance of determining sample size based on the number of variables and complexity of framework, not the size of population. Ho (2006) asserted that sample size is one of the requirements in the checklist for SEM. He revealed that in SEM, sample size must be sufficiently large. Although there is a lack of agreement on the meaning of "sufficiently large", the most appropriate minimum ratio is ten respondents per parameter, which increases proportionately as the complexity of the model increases (Ho, 2006).

Different rules have been recommended for determining the required sample size for reliable factor analysis in SEM. The most accepted rules suggest determining sample size based on the number of variables being analysed. According to Schumacker & Lomax (2010) the sample size could range from ten subjects to twenty subjects.

According to Byrne (2010) small sample size tends to over-reject true population models. Kline (2008) noted that small sample size is expected to decrease the power of statistical tests. Both authors mentioned that a sample size of less than 100 cases may be too small, between 150 and 200 cases is a better minimum, but perhaps even 200 cases is insufficient to analyse more complex models, which require the estimation of more parameters. Thus, 20 cases for every estimated variable is ideal, although 10 subjects per variable is less ideal, but may suffice (Schumacker & Lomax, 2010).

At a certain alpha level, rising sample size creates a larger power of statistical test and causes a problem of too much power. Too much power means that by raising the sample size, smaller effects would found to be statistically significant, and then at extremely big sample sizes almost any effect is significant. Accordingly, the number of subjects can affect the statistical test by either making it insensitive at small sample sizes or overly sensitive at very large sample size. Thus, for sample sizes that exceed 400 subjects, which is considered a big sample size, the researcher should be cautious of too much power and look at all significant results to make sure that the results have practical significance as well (Hair, Black, Babin, Anderson, & Tatham, 2006).

In this study, as suggested in the mentioned resource (Schumacker & Lomax, 2010), the highest logical number of subjects per variable is considered due to the complexity of framework. Thus, the number of respondents was decided based on multiplying the 12 variables in addition to three dimensions of achievement motivation by 20 subjects, which is equal to 300. Also, according to Ary et al. (2002), if response rate remains about 75 percent, the researcher has to think about the non-respondents. Thus, by planning on follow-up for improving the response rate, the possible rate of non-response is considered 25%, and that figure is added to the total sample size. Finally, based on the problem of too much power in large sample sizes, the decided sample size was 375.

Thus, 375 questionnaires were distributed, and after all the follow-ups, the number of collected questionnaires was 303. To check the adequacy of sample size, the G*Power software with multiple linear regression analysis was used by considering effect size of 0.15, α err probability of 0.01, power of 0.99 and 12 predictors. The total required sample size is calculated as 295 cases, which confirmed the adequacy of the collected questionnaires.

Instrument Development

The questionnaire for this study includes 4 sections. Since there is no available questionnaire for achievement motivation, and the limited research in this area has taken place only among young subjects for predicting their future career achievement motivation, the instrument development of this research involves modifications and adaptations. Further, it is important to note that a general modification in most of the scales, which will be discussed later, considers a similar response range from 1 to 5. According to Hair, Black, Babin, and Anderson (2010), there are two reasons for making a similar response range, which are: first, scales that contain more than four response categories can be treated as the interval scale, which is a better scale in comparison to the ordinal scale for using SEM. Second, all the items indicating a factor need to be normalized and mathematically transformed to a common scale range prior to using SEM.

Section A consists of questions related to the background factor variables. The background factor variables include career stage, gender, gender identity, leadership stereotypes, and work experience, in which gender is nominal, and age and work experience are continuous variables. Gender identity is assessed using Short Bem Sex-Role Inventory (BSRI) cited in (Hensley, 2002; Powell & Butterfield, 2003). The review of literature on leadership stereotypes had shown that there is no standard inventory developed by previous researchers to measure this variable. However, using reviewed literature, three sources of information were chosen to develop the scale. First, the research reported by Powell, Butterfield, and Parent (2002), which is a longitudinal study about leader stereotypes; then a scale for two measures related to attitudes toward males in society from Falkenbert, Hindman, and Masey, (1983); and a scale of attitudes toward women in society from Spence and Helmreich, (1978) which was adapted by Hensley (2002).

Section B consists of scales for measurement of three dimensions of achievement motivation, including aspiration, mastery, and career salience. Aspiration is measured by the coping version of career aspiration scale of O'Brien Gray, Tourajdiyhjr4, and Eigenbrode (1996) (cited in Cannon, 2004; Hensley, 2002; and Leal, 2006) that was translated to Persian and adapted in Karami et al., (2009, 2011). The edited version of the Work and Family Orientation Questionnaire (WOFO) designed by Helmreich and Spence (1978) cited in Leal (2006) was used in this study to measure the mastery motivation. The latest version of salience scale developed by Farmer (1997) was used to assess the respondents' salience toward achievement motivation in leadership role.

Section C is related to personal factor variables that consist of self-efficacy, self-esteem, achievement value, and extension agents' leadership competencies. Self-efficacy is assessed with the General Self-Efficacy Scale (GSE) designed for the adult population (Burke, 2005). General self-esteem was measured using items designed by Rosenberg (1965). Consistent with recent research (Ladebo et al., 2008) that used this scale among extension agents and reported the internal consistency reliability (Cronbach alpha) of 0.72, the achievement value consists of subjective and objective value. The measure for the subjective value was developed by the researcher according to reviewed literature, which consists of four items. The Work Preference Inventory (WPI) is the basis of intrinsic and extrinsic achievement motivation value (two dimensions of objective achievement value), which was developed by Amabile, Hill, Hennessey, and Tighe (1994). Extension agents' leadership competencies scale was established through reviewing the literature, such as Khalil et al. (2008), Burke (2002) and Ladewig and Rohs (2000). The scale consists of 9 aspects of leadership competency for extension agents: problem solving, coaching, open communication and sharing of knowledge, volunteering to train others, organizational skills, team goals, personal-interaction skills, verbal communication, and written communication (Burke, 2002). Also the definition of leadership competencies in leadership development program of extension agents includes managing the job, building the team, thinking clearly, and relating to others (Ladewig & Rohs, 2000).

Section D covers the environmental factor variable measurements, involving mentoring, extension organization supports, and extension organization supervisors' support variables. To measure mentoring, an adapted version of the scale developed by Ricciardi (2005) was used. The scale of Perceived Organizational Support cited in Hochwarter et al. (2003) was used to measure organizational support. The scale of supervisors supports cited in Xu (2007) from Maurer, Weiss and Barbeite, (2003) with a reliability of .92, was used to measure Extension organization supervisors' supports.

Validity and Reliability

Reliability and validity are two fundamental issues in all measurements, concerning how existing measures are associated to the constructs. Validity refers to the way a researcher abstracts the idea in a conceptual definition and a measure (Neuman, 2007). Validity has three distinct aspects, including criterion, construct, and content validity. The criterion validity investigates the agreement of scale with an external source, including pre-existing research (concurrent validity) or future behaviour (predictive validity) (Muijs, 2004). The criterion validity is mostly applicable on the scale that is developed by the researcher. The scales of present research were mostly pre-existing, which were tested in other research and show an appropriate level of criterion validity.

The construct validity is related to the internal structure of an instrument and the concept it is measuring. The assessment of construct validity is one of the advantages of using SEM (Muijs, 2004). Construct Validity could be described through convergent and discriminant validity. Convergent validity refers to the proportion of share variance by the items that are the indicators of a specific construct. The convergent validity could be assessed by considering the size of factor loading on their underlying construct, which should be .5 or higher. Also, items of measure need to be significant on their underlying constructs to be considered to have convergent validity. Another indicator of convergent validity is Average Variance Extracted (AVE) among a set of a construct's items that should be equal to or in excess of .5. AVE is calculated by dividing the sum square of standardized factor loading by the number of factor loadings. Discriminant validity shows how well a construct is truly distinct from another construct. The square of covariance between each two constructs should be less than AVE of each one of the constructs to support their discriminant validity (Hair et al., 2010). Construct validity could be established after theoretical testing of an instrument, thus, its related results are provided in chapter four.

Content validity refers to consistency of scale, theoretical definition of concept, and how that concept works. The face validity of measure is also important, which is related to the judgment of whether the instrument looks good and appropriate. Usually a panel of experts who know the theoretical foundation of a concept judge the face and content validity of a scale (Muijs, 2004). Therefore, content and face validity of an instrument, prior to conducting the back translation procedure, were improved by passing to supervisor and committee members. After back translation, an Iranian subject matter specialist was consulted to improve the face validity of the Persian version of the instrument.

The questionnaire was prepared in Persian, which is the original language of the respondents. Thus, to check the validity of the scale, which was mostly in English, translation and back translation were implemented. As Farmer, in her permission letter suggested, a person highly competent in both Persian and English was chosen to translate the measure from English to Persian. For this purpose one Iranian PhD student majoring in Teaching English as a Second Language (TESL) was chosen. A different person with the same characteristics and highly competent in both Persian and English was chosen to translate the Persian version back into English. Through the comparison of the first English version with the second one, the researcher corrected the inconsistencies.

To assess the reliability and internal consistency of the data, a pilot study was conducted. The pilot test was conducted in the Administration of Jihad-e-Agriculture in Bukan County and its related Agricultural Extension and Services Centres Located in districts, sub-districts, and villages from Azarbayejan Gharbi province. The permission to conduct the pilot test was granted through a meeting with the head of Administration of Jihad-e-Agriculture in Bukan County. The method was drop-and-pick, and the number of questionnaires completed for pilot test totalled 32 cases. As shown in Table 6-1, all variables display acceptable values of Cronbach's alpha. The reported value of aspiration is slightly questionable; although it has the potential of improving by deleting one item, the item was kept due to its importance and later in real data collection shows no sign of any problem.

Table 6-1: Results of reliability of variables in pilot test

Variables	Original Reliability	Cronbach's Alpha in Pilot Test
Mentoring	.89	.870
Organizational support	.92	.924
Supervisors supports	.92	.951
Self-esteem	0.72	.731**
Self-efficacy	.80	.908
Leadership competencies	-	.939
Intrinsic achievement value	.77	.708
Extrinsic achievement value	.64	. 755
Subjective achievement value	-	.964
Leadership stereotypes	-	.814
Gender identity	.775	.787
Aspiration	.877	.682*
Mastery	.61	.712*
Salience	.83	.737

*After deleting one item, **After deleting two items

Data Collection

The data of study were collected using questionnaires. Permission to conduct the study among extension agents in three provinces was granted through a letter asking to introduce the researcher and the research to the manager of staff training in Ministry of Jihad-e-Agriculture. The researcher was presented in the three selected provinces and submitted the letters to the heads of extension offices in the provinces. Then, to perform the new procedure of questionnaires confirmation, as explained by the head of the extension office, the questionnaires were submitted to the preservation office of the province. As a result, the first item of organizational support scale “My organization considers my goals and values” was deleted and some of the wording was changed. Then the researcher, in a meeting with the head of the extension office in the provinces, asked for support in terms of: getting the respondents’ lists, having encouragement for the respondent’s participation in the study, as well as how to contact the extension office in counties and Agricultural Extension and Services Centres Located in districts, sub-districts, and villages.

The procedure of data collection was drop and collect in the provinces, and postal questionnaires in counties and centres. The questionnaires were distributed in provinces by the researcher and the questionnaire for counties and centres of each province were posted along with a letter from the heads of extension offices in the related province. In the first stage the respondents were given 2 weeks to complete the questionnaires.

The researcher tried to keep in touch through phone calls with the provinces, counties, and centres for follow-up and to answer possible questions or misunderstandings about the data collection process. Also the researcher’s email was given in the questionnaire’s introduction for possible questions or suggestions, which was another way to keep in touch with respondents.

To increase the response rate, in addition to adding possible non-response rates to the calculating sample size (25%), and the follow-up procedures, the researcher tried to encourage the heads of extension offices in the provinces to give more support by sending the second and third follow-up letters to the counties and centres, or through phone calls. In doing so, the researcher promised to analyse the data of each province separately and to write the report of results in Persian for each province separately. The Persian report of the analysed data in each province allowed ease of use for the readers, who are mostly not familiar with

English, provided a good opportunity for the head of extension office of the province to get the update and reliable information, which will be useful in their decision making and in formulating organizational strategies. Finally, from 375 distributed questionnaires, 303 questionnaires were collected and used for analysis.

Confidentiality of the respondents in the study was protected by having no name or identification codes written on the questionnaires, therefore, there was no way of matching respondents' names with their responses.

Table 6-2 shows the data collected from three Organizations of Jihad-e-Agriculture in the provinces, namely Tehran, Kermanshah, and Khorasan Razavi, which are the main clusters of the study. All together 53 Counties, and 123 Agricultural Extension and Services Centres were in the sub-clusters of the study.

Table 6-2: The distribution of study clusters and questionnaires

Province	Counties	Centers	Distributed Questionnaires	Collected Questionnaires
Tehran	13	25	140	108
Kermanshah	14	51	120	103
Khorasan Razavi	26	47	115	92

Data Analysis

Descriptive and inferential statistics were used to analyse the data. Descriptive statistics were used to summarize the response for all independent and dependent variables. Frequency measures such as percentages, mean, and standard deviation were used at this stage.

The factor analysis and path analysis were used as the inferential statistics through Structural Equation Modelling (SEM). The SEM was used in this study because it was a more suitable approach to analyse the data for which the inter-variable relations have been already specified based on the theory. Also, dissimilar to most other multivariate methods which are fundamentally descriptive, SEM tests hypothesis and takes a confirmatory rather than an exploratory approach to the data analysis which is required in this research. Furthermore, while conventional multivariate methods are incapable of either assessing or correcting for measurement error, SEM produces clear estimates of these errors, which, in research with established measures similar to this research, is important. Lastly, while the traditional multivariate methods are based on observed measurements only, SEM can incorporate both unobserved (i.e., latent) and observed variables. Thus SEM is utilized successfully for tasks that deal with numerous research questions involving non-experimental research (Byrne, 2010). Further, Schumacker and Lomax (2010) consistent with the above, discussed the reasons to conduct SEM over common methods, noting that they provide an opportunity for looking at the multiple observed variables to better understand the area of scientific inquiry, give greater recognition to the validity and reliability of observed score by simultaneous consideration of measurement error and statistical analysis of data, have the ability to analyse more advanced theoretical models, and lastly have become increasingly user friendly and easy to use.

Structural Equation Modelling (SEM) is a theory based statistical methodology. The term structural equation modelling carries two key aspects of the method: First, the causal processes under study are represented by a series of structural equations; and second, these structural equations can be modelled graphically to make possible a clearer conceptualization of the theory under study. The hypothesized model could then be tested statistically in a simultaneous analysis of the entire system of variables to determine the extent to which it is consistent with the data. In using factor analysis, the researcher examined the co-variation among a set of observed variables in order to gather information on their underlying latent constructs (Byrne, 2010). Confirmatory factor analysis (CFA) was properly used when the researcher utilized strong theories and empirical bases that allow the researcher to specify an exact factor model in advance (Byrne, 2010; Stevens, 2002), such as in the present study. Prior to conducting the SEM, the data should be screened in terms of missing data and outliers, which are the subsequent subsections of data analysis. Then, the Goodness-Of-Fit indices would be discussed, which are the criteria for assessing model fit in measurement and structural models. Conducting SEM according to its two key aspects produces two models: the measurement model and the structural equation model that will be presented in the next subsections. Furthermore, in the measurement model, assessment of normality, multicollinearity, and construct validity are of importance, such that each is discussed as a subsection of data analysis after measurement model. The test of mediating effect was presented as the last subsection.

Missing Data

Detecting the missing data is one of the important issues that needs to be considered in every analysis. There are different options for handling the missing data, including deleting subjects who have missing data, using Pairwise or Listwise deletion, replacing the missing data using mean or regression imputation, or using a statistical procedure that accommodates the presence of missing data. Using SEM, regression imputation is one reliable way in which the missing value would be predicted based on other variables which are not missing (Byrne, 2010). Thus, in this study the missing data were detected using the regression imputations method, although the number of missing data was small.

Outlier

Outliers refers to data values that are extreme (Schumacker & Lomax, 2010) and could influence the mean, standard deviation, and correlation coefficient values. The sources of outliers vary from data entry, to observation error, measurement error, or a respondent's extreme point of view (Schumacker & Lomax, 2010). The Mahalanobis distance (d-squared) measure is the test to check outliers and is defined as "the distance in standard deviation units between a set of scores for one case and the sample mean (centroid)" (Hair, Black, Babin, Anderson, & Tatham, 2006, p. 65). According to Byren (2010) a d-squared value that stands distinctively apart from all other d-squared values shows the possibility of an outlier. Further, if the d-squared divided by the degrees of freedom value exceeds 2.5 in small samples (n<100) and exceeds 3 or 4 in large samples, this shows the possibility of an outlier (Byrne, 2010; Hair, et al., 2010). In the present study, the Mahalanobis distance (d-squared/degrees of freedom) value was used as the criterion to find outliers, and the results showed that there was no potential of outliers in the dataset.

Goodness-Of-Fit (GOF) Measures

The validity assessment of measurement and structural model depends on the value of Goodness-Of-Fit (Hair et al., 2010). Goodness-Of-Fit (GOF) measures are the extent to which the observed covariance input matrix corresponds with or departs from that predicted by the proposed model (Ho, 2006). GOF measures, such as Chi-Square, which is one of the long-standing indicators of overall goodness-of-fit, is sensitive to sample size. It means a large sample may indicate a significant Chi-Square when it should not have been significant (Hair et al., 2010). Also, according to Hair et al. (2010) and Ho (2006) the Goodness-of-Fit Indicator (GFI) and the Adjusted Goodness-of-Fit Indicator (AGFI) generally have the possible range from 0 to 1, and values greater than .9 indicates better fit and a meaningful model, even when the Chi-Square is significant. Another measure is the Root Mean Square Error of Approximation (RMSEA), in which a measure greater than .1 indicates a poor fit, values ranging from 0.08 to 0.1 indicate mediocre fit, and values ranging between 0.03 and 0.08 indicate better model fit. Further, the Comparative Fit Index (CFI), Normed Fit Index (NFI), and Tucker Lewis Index (TLI) each indicate a good fit to the model at about .9 or greater, with 1 indicating a perfect fit for the model (Hair et al., 2010; Ho, 2006). Jaccard and Wan (1996), cited in Garson (2009), recommended the use of at least three fit tests to reflect diverse criteria. Kline (2008), cited in Garson (2009), recommends at least four tests, including chi-square; GFI, NFI, or CFI; NNFI; and SRMR. Moreover, Hair et al. (2006) suggest that reporting X^2 value and degrees of freedom, along with CFI and RMSEA, often provides sufficient unique information to evaluate. On the other hand, according to Hair et al. (2006), when three to four indices, including one of the absolute indices and one of the incremental indices, meet the criteria, it provides adequate evidence of model fit. The χ^2 and RMSEA are absolutes fit indexes, and incremental fit indices include TLI, NFI, IFI, and CFI.

The Measurement Model

The measurement models specify how the latent variables are measured in terms of the observed variables and are concerned with the relations between observed and latent variables (Ho, 2006). In this model, the measurement properties of observed variable are described which specify the hypothesized relations between a set of observed variables, such as questionnaire items, and the unobserved variable or construct they were designed to measure.

The measurement model is important due to the fact that it provides testing for the factorial validity of scores from a measuring instrument. In this step the researcher sought to determine the extent to which items were designed to measure a particular factor or latent construct. In general, sub-scales of a measuring instrument are considered to represent the factors; all items comprising a particular sub-scale therefore are expected to load onto their related factor. However, to measure constructs by multi-item scales, parcels were used as indicators of latent variables. Parcels are aggregations (sums or averages) of several individual items (Coffman & MacCallum, 2005). Parcelling provides a way of dealing with an unmanageable number of measured reflector variables per construct (Hair et al., 2006). Parcelling has the potential to improve model fit because it reduces the complexity of the model, and models with fewer variables have the potential for better fit (Hair et al., 2006). Another advantage of using parcels as indicators of constructs is that parcels generally have higher reliability than single items (Coffman & MacCallum, 2005).

To design the measurement model, the items of each scale are specified. Then, each latent construct to be included in the model is identified and the measured indicator variables (items) are assigned to a latent construct (Hair et al., 2006). To measure constructs by multi-item scales, parcels were used as indicators of latent variables. As revealed by Coffman & MacCallum (2005), it is better to use parcels to serve as indicators of latent variables rather than using total scale scores, such as a path analysis model. Further, in multi-item scales and scales with more than 15 items, parcelling is advised (Hair et al., 2006). Therefore, the study used parcels in multi-item scales including supervisor supports, organizational supports, and leadership competency. Supervisor supports were measured by two sub-scales, including work support with 10 items, and non-work support with 4 items, and were indicated in the measurement model with two parcels of work and non-work support.

Similarly, the organizational supports were measured by two sub-scales, including work supports and non-work supports, and were indicated in the measurement model with two parcels of work and non-work supports. Another variable indicated with parcels was leadership competency, which was measured by 18 items. According to Hair et al. (2006), parcelling is applicable when a large number of items being used to measure one construct in which they should have a reliability of .9 or better. Thus, parcelling was appropriate for leadership competency construct, which was measured using 18 items and had a very good reliability of .945. Since this variable had no sub-scale, the factor analysis was conducted to distinguish how items were classified into the factors. Then, based on the factor analysis results, the items were classified in two factors; the first factor with 10 items and the second factor with 8 items. The average scores of items in each factor were calculated as a parcel.

The initial measurement model of the study failed to meet the required GOF indices in the confirmatory test of the model: $\chi^2(3491)$ =7250.768, p= .000 χ^2/DF=2.07; GFI = .649, AGFI =.624, CFI = .716, NFI=.571, IFI = .719, TLI = .702, RMSEA = .060.

Thus, several solutions such as investigations on factor loadings and modification indices to remove the poor items were estimated and tested. The adequacy of factor loadings were assessed according to the Hair et al. (2010) and Kline (2011) rules that judge the size of factor loading higher than .5 as a criterion to determine the construct validity of the measurement.

Thus the eliminated items based on the mentioned criteria are as follows: one item of mentoring, four items of self-efficacy, five items of leadership stereotypes, three items of intrinsic achievement value, six items of extrinsic achievement value, one item of subjective achievement value, three items of aspiration, two items of mastery, and five items of salience. Further, according to Farmer (1997), a variable with low construct validity could consider to drop out of the framework. Thus, to get the best model fit, two variables which showed low construct validity (convergence validity) and high modification indices with other variable were dropped. The self-esteem and organizational support were the dropped variables. Theoretically, self-esteem was used in academic achievement, thus, it is justifiable to drop such a variable with low construct validity. Organizational support was another dropped variable and perhaps the reasons to drop this variable lie in the respondents' biases.

Table 6-3 shows the GOF indices of the final measurement model (Figure 3-4). The indices show that the measurement model presents a relatively good fit between the data and the proposed measurement model.

χ^2 (1072) =1570.416, p= .000 χ^2/DF=1.465; GFI = .831, CFI = .923, IFI = .924, TLI = .916, RMSEA = .039.

It could be interpreted that CFI, IFI and TLI significantly pass the cut-off value (0.9). In addition, the RMSEA was 0.039, which falls between the recommended range of acceptability (between 0.03 and 0.08) according to Hair et al. (2006). Thus, the measurement model shows a good fit with the data of the study.

Table 6-3: GOF indices of the measurement model

Index	CMIN	GFI	CFI	IFI	TLI	RMSEA
Value	1570.46 (P=.00)	.831	.923	.924	.916	.039

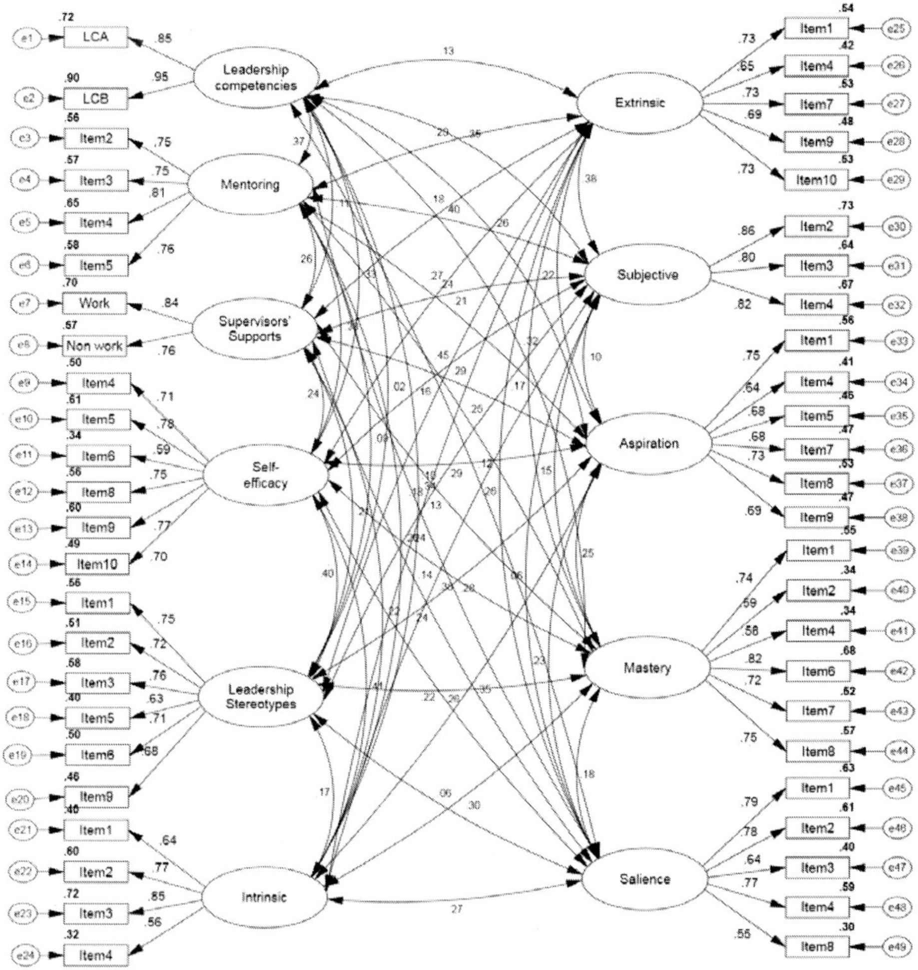

Figure 6-4: The measurement model of the study showing the relationships of items and constructs

Table 6-4: The results of CFA for measurement model

Constructs and Indicators (items/parcels)	Standardized Factor Loading	Construct Reliability	AVE (>0.5)
[Background Factors]			
Leadership stereotypes		.857	.50
- Item 1	.717		
- Item 2	.746		
- Item 3	.759		
- Item 5	.629		
- Item 6	.708		
- Item 9	.682		
[Personal Factors]			
Self-efficacy		.862	.52
- Item 4	.780		
- Item 5	.709		
- Item 6	.587		
- Item 8	.751		
- Item 9	.773		
- Item 10	.699		
Subjective achievement value		.865	.68
- Item 2	.802		
- Item 3	.818		
- Item 4	.856		
Extrinsic achievement value		.832	.50
- Item 1	.735		
- Item 4	.648		
- Item 7	.726		
- Item 9	.695		
- Item 10	.728		
Intrinsic achievement value		.791	.51
- Item 1	.635		
- Item 2	.774		
- Item 3	.847		
- Item 4	.565		
Leadership competencies		.945	.81
- Factor 1items	.851		
- Factor 2 items	.949		

Constructs and Indicators (items/parcels)	Standardized Factor Loading	Construct Reliability	AVE (>0.5)
[environmental Factors]			
Supervisors supports		.940	.64
- Work items	.758		
- Non work items	.838		
Mentoring		.852	.59
- Item 2	.747		
- Item 3	.754		
- Item 4	.808		
- Item 5	.765		
[Achievement Motivation]			
Aspiration		.848	.50
- Item 1			
- Item 4	.750		
- Item 5	.637		
- Item 7	.680		
- Item 8	.729		
- Item 9	.688		
Mastery		.851	.50
- Item 1	.744		
- Item 2	.585		
- Item 4	.581		
- Item 6	.823		
- Item 7	.723		
- Item 8	.754		
Salience		.830	.51
- Item 1	.794		
- Item 2	.781		
- Item 3	.636		
- Item 4	.770		
- Item 8	.552		

Normality

Prior to conducting the structural model the data were subjected to tests of normality in the measurement model through checking both the symmetry (skewness) and the "flatness" (kurtosis) in meeting the acceptable assumption. According to Kline (2011), if the value of the skew and kurtosis fall within a reasonable range, it would satisfy the multivariate normality assumption. The importance of normality assessment is due to using Maximum Likelihood Estimates (MLE) in the present study. The non-normal data will affect variance-covariance among variables and can occur due to a limited size of sample or scaling of variables such as using ordinal rather than interval scale (Schumacker & Lomax, 2010).

According to Schumacker and Lomax (2004; 2010) the data may be assumed to be normal if skewness and kurtosis is within the range of (-1 to +1, or -1.5 to +1.5), while Garson (2009) indicated the range of +/- 1.0 (some stated +/- 1.5 or even 2.0). The result of normality assessment in measurement model based on skewness and kurtosis criteria falls within the range of +/- 1.0 which reveals no item to be non-normal. Skewness ranged from -1.037 to .094 and kurtosis ranged from -1.088 to .672, which meet the required values.

Multicollinearity

Another key consideration is the issue of multicollinearity that arises when two or more variables are so highly correlated (Alareqe, Mohammed, Harun, Samah, & Alameary, 2008; Byrne, 2010). Multicollinearity shows that two or more highly correlated variables essentially represent the same underlying construct (Byren, 2010; p, 168). Hair et al. (2006) stated that a multiple correlation of .9 between the independent variable and all others would result in a tolerance value of .19, and hence a possible problem of multicollinearity. In the present measurement model the results of correlation estimates among the constructs ranged from .06 to .45. Thus, there were no problems of multicollinearity.

Construct Validity

In the measurement model the assessment of validity should also be considered. Historically, validity is defined as the extent to which an instrument has measured what it claimed to measure. Recently, validity is defined as the degree to which evidence and theory support the interpretations of test scores entailed by proposed uses of tests (Ary et al., 2002). There are different kinds of validity which should be considered in SEM approach, such as convergent validity and discriminant validity, which are representative of construct validity. Construct validity is the extent to which a set of measured items actually reflect the theoretical latent construct which was satisfied in this study. Convergent validity, as a part of construct validation, is concerned with the items that are indicators of a specific construct which should share a high proportion of variance. Convergent validity was assessed using three indicators, including standardized factor loading, Average Variance Extracted (AVE), and construct reliability. The factor loading needs to be significant with a value higher than .5; the average variance extracted among the set of construct items should be higher than .5; and construct reliability uses the same rule as Cronbach's alpha in which the value of .7 is considered as acceptable, .8 good, and .9 shows a very good level of construct reliability (Hair, et al., 2010). Furthermore, construct reliability could be calculated using the following formula recommended by Hair et al. (2006):

$$CR \quad \frac{[\sum^{n}_{i=1} \lambda_i]^2}{[\sum^{n}_{i=1} \lambda_i]^2 + [\sum^{n}_{i=1} \delta_i]}$$

Where;

λ_i =Factor loading

δ_i = error variance

The results show that first, all the items or parcels had a high standardized factor loading on their underlying constructs (values ranged from .552 to .949) (Figure 6-4). In addition, all the items on their underlying constructs were significant at 0.001. Second, the convergent validity of constructs was assessed based on the AVE. The AVE for all constructs exceeded the minimum criterion of .50, indicating that the majority of the variance was explained by the constructs (Hair et al., 2010) (Table 6-4). As shown in Table 3-4, the value of constructs' reliability ranged from .791 to .945, which shows each construct's reliability exceeded the minimum requirement of 0.7, ensuring adequate internal consistency (Table 6-4).

Discriminant validity shows how well a construct is truly distinct from another construct. The square of covariance between each two constructs should be less than the AVE of each one of the constructs to support their discriminant validity (Hair et al., 2010). Discriminant validity was also met as covariance among the constructs range from .06 to .45 (see Figure 6-4) which shows that their squares were less than the lowest AVE.

Structural Equation Model

The second part of conducting SEM is structural equation modelling, which is a flexible, comprehensive model that specifies the pattern of relationships among independent and dependent variables, either observed or latent, along with the indirect effects (Ho, 2006). The structural equation model is an inclusive model that describes the configuration of independent and dependent variable relationships, either observed or latent

(Hair et al., 2010; Ho, 2006; Landis, Beal, & Tesluk, 2000). It incorporates the strengths of multiple regression analysis, factor analysis, and multivariate ANOVA (MANOVA) in a single model that can be evaluated statistically. Moreover, it offers a direct test of the theory of interest through directional predictions among a set of independent or a set of dependent variables, and modelling of indirect effects (Ho, 2006). The confirmed items and variables in the measurement model, along with the research framework, which is provided according to the theoretical model, are the bases for structural equation model development.

Test of Mediation Effect

The mediation effects of personal and environmental factors were assessed through path analysis. The antecedents' conditions of background factors have a relationship with the criterion of achievement motivation by an intervening mechanism of personal and environmental factors. Thus, the casual chain of the above mentioned factors were assessed through path analysis using normal methods of mediation testing, as explained by Baron and Kenny (1986) and Mallinckrodt et al. (2006). This method is called the basic normal theory approach, in which four necessary conditions are applied. First, there must be a significant correlation between the predictor variable and the dependent variable. Second, the predictor variable must account for a significant proportion of the variance in the mediating variable. Third, the mediator variable must account for a significant proportion of variance in the dependent variable. Fourth, the indirect association between the predictor variable and the dependent variable must be significantly less than the direct association between the predictor variable and the dependent variable (Baron & Kenny, 1986; Mallinckrodt et al., 2006; Mathieu & Taylor, 2006). However, Kenny et al. (1998) cited in Mallinckrodt et al. (2006), argued that the first condition is not necessary. The mediation effects of each variable of background factors were assessed using the aforementioned conditions.

Also, mediating effect significance of personal and environmental factors were determined through the new approach of "bootstrapping". Bootstrapping provides the most powerful and reasonable method of obtaining confidence limits for specific indirect effects under most conditions (Bauer, Preacher, & Gil, 2006; Preacher & Hayes, 2008). Bootstrap methods provide an estimate of the magnitude of the indirect effect, its statistical significance, and determine confidence intervals for the point estimate (Mallinckrodt et al., 2006). It is important to note that bootstrapping did not provide the indirect effects of each specific variable, which, in this study, were covered through the normal approach of Baron and Kenny.

Operational Definitions

Achievement motivation in leadership role: Achievement motivation in leadership role was considered as a desire or interest for attainment of excellence in leadership role for achieving improved performance, and is composed of three dimensions aspiration, mastery and salience.

Aspiration: Aspiration is considered as a mental desired level of leadership role.

Mastery: Mastery is considered as desire for intellectual challenge; desire to choose difficult, challenging, but achievable tasks, and to keep struggling to master the task of leadership role.

Salience: Salience is considered as the extent to which a person sees involvement in leadership role as central to his or her adult life role.

Emergent leadership role: Emergent leadership role of extension agents refers to encouraging clients in a community in order to bring about their empowerment and that of society following an appropriate approach in a democratic way by considering that it is an achievable role; not because of a formal position in an organization, but because of the way clients respond to extension agents.

Gender identity: Gender identity refers to the sex–appropriate preferences or characteristics an individual develops or learns rather than characteristics that appear from biological origins, including masculine and feminine traits. Masculine traits are those that are believed to be highly possessed by males, such as independence and dominance. Feminine traits, such as gentleness and sensitivity to the feelings of others, are mostly believed to be possessed by females.

Leadership stereotypes: Leadership stereotypes refer to a set of fixed ideas or beliefs about the personal attributes of leadership role which may not really be true.

Self-esteem: Self-esteem refers to individual judgment of self-worth and confidence to handle a job role, which in this case is leadership.

Self-efficacy: Self-efficacy refers to individuals' beliefs about their capability to successfully perform the given task of a leadership role.

Achievement value: Achievement value refers to the individual's subjective and objective value underling achievement of leadership role.

Subjective achievement value: Subjective achievement value refers to persistence in importance of individual and work-family balance in achieving leadership role.

Objective achievement value: Objective achievement value refers to persistence in importance of personal interests in achieving leadership role, including receiving more pay, higher position, and advancing at a faster rate (extrinsic values), or exercising competencies and skills and making meaningful contributions to society (intrinsic value).

Intrinsic value: Intrinsic value refers to the importance of achieving a leadership role as an opportunity to exercise competencies, skills, and make meaningful contributions to the society.

Extrinsic value: Extrinsic value refers to the importance of achieving leadership role to attain something distinct from the work itself, such as financial success, having control and influence over others, and occupying a prestigious position at work.

Extension agents' leadership competencies: Extension agents' leadership competencies are described as the knowledge, abilities, and skills used by extension workers to guide the clients.

Mentoring: Mentoring refers to a teaching and learning process in which an experienced employee serves as a role model to the less experienced employee in terms of career functions.

Extension organization supports: Extension organization supports refer to extension an agent's perception of the extension organization's concern about their well-being, which consists of work supports and non-work supports. The work supports are related to the organizational structure that supports work goal attainment, along with professionalism, developmental opportunities, and meritocracy. The non-work supports refer to a family friendly organizational structure that aims to help employees by providing a balance between work and family.

Extension organization supervisor supports: Extension organization supervisor supports refer to the helpful role of extension supervisors as crucial organization agents in creating a supportive work and non-work environment.

Conclusion

This chapter provides discussion of the methodological procedures of research, defining the research paradigm as quantitative, research design as cross-sectional, and research method as survey. Also the research framework and related objectives and hypotheses were presented. Then, by introducing the population of study, which is scattered over a wide geographical region, the type of sampling was decided as cluster sampling. The sample size was determined based on the number of predictor variables and complexity of framework as suggested by SEM scholars, and the adequacy of the required size of sample, checked by G*Power software. The process of questionnaire development and investigation of its validity and reliability was discussed as the study instrument. Lastly, the data collection and data analysis were explained. The data analysis was discussed in more detail to provide the motivation for using SEM, how to handle the missing data, and explained the measurement model, outliers, the normality, the multicollinearity, and the construct validity, then the structural model and related GOF indices were discussed along with the assessment of mediation effects. The operational definitions have also been provided at the end of this chapter.

References

Alareqe, N. A. H., Mohammed, O. B., Harun, L., Samah, B. A., & Alameary, A. A. (2008). *Relationship between Belief System and Depression with Anxiety among Undergraduate Students in Yemen.* Paper presented at the Asian Cognitive Behavior Therapy (CBT) Conference, October18 to 21, Bangkok, Thailand.

Amabile, T. M., Hill, K. G., Hennessey, B. A., & Tighe, E. M. (1994). The work preference inventory: assessing intrinsic and extrinsic motivational orientations. *Journal of Personality and Social Psychology, 66*(5), 950-967.

Ary, D., Jacobs, L. C., & Razavieh, A. (2002). *Introduction to research in education* (sixth ed.). United States of America: Wadsworth, Thomson Learning.

Baron, R. M., & Kenny, D. A. (1986). The moderator-mediator variable distinction in social psychological research: conceptual, strategic, and statistical considerations. *Journal of Personality and Social Psychology, 51*(6), 1173-1182.

Bauer, D. J., Preacher, K. J., & Gil, K. M. (2006). Conceptualizing and testing random indirect effects and moderated mediation in multilevel models: new procedures and recommendations. *Psychological Methods, 11*(2), 142-163.

Burke, M. U. (2005). *Family influence on self-efficacy: exploring the relationship between perceived parenting style, current social support, and self-efficacy belief in a sample of nontraditional college students (doctoral dissertation)* PhD, Capella University, Minnesota.

Burke, T. B. (2002). *Defining competency and reviewing factors that may impact knowledge, perceived importance and use of competencies in the 4-H professional's job (doctoral dissertation).* Ph.D, North Carolina University Raleigh.

Byrne, B. M. (2010). *Structural equation modeling with AMOS: basic concepts, applications, and programming* (second ed.). New York, London: Taylor and Francis Group.

Cannon, S. C. (2004). *Supporting first-year college women in math, science, and related majors: A Career Development Intervention (Doctoral dissertation).* Doctor of philosophy, North Carolina State University, United States.

Coffman, D. L., & MacCallum, R. C. (2005). Using parcels to convert path analysis models into latent variable models. *Multivariate Behavioral Research, 40*(2), 235-259.

Farmer, H. S. (1997). Women's motivation related to mastery, career salience, and career aspiration: A multivariate model focusing on the effects of sex role socialization. *Journal of Career Assessment, 5*(2), 355-381.

Garson, D. (2009). Structural Equation Modeling, Retrieved 1 December, 2009, from http://faculty.chass.ncsu.edu/garson/PA765/structur.htm

Hair, J. F., Black, W. C., Babin, B. J., & Anderson, R. E. (2010). *Multivariate data analysis*. United States of America, Pearson prentice hall.

Hair, J. F., Black, W. C., Babin, B. J., Anderson, R. E., & Tatham, R. L. (2006). *Multivariate Data Analysis*: Upper Saddle River, NJ: Prentice Hall.

Henn, M., Weinstein, M., & Foard, N. (2006). *A short introduction to social research*. London, thousand, Oaks, New Delhi: SAGE press.

Hensley, L. (2002). *The Influence of Gender Role and Gender Stereotyping on the Career Choice and Career Commitment of Adolescents*. Retrieved 29 Jan, 2009 http://www. samford.edu/schools/artsci/scs/hensley.html.

Ho, R. (2006). *Handbook of univariate and multivariate data analysis and interpretation with SPSS*. Boca, Raton, New york: Chapman & Hall/CRC ,Taylor and Francis Group.

Hochwarter, W. A., Kacmar, C., Perrewe, P. L., & Johnson, D. (2003). Perceived organizational support as a mediator of the relationship between politics perceptions and work outcomes. *Journal of Vocational Behavior, 63* 438-456.

Holmes, R., Dahan, H. M., & Ashari, H. (2006). *A guide to research in the social sciences*. KL, Malaysia: Pearson Prentice Hall.

Iran-Statistics-Center. (2009). *Statistical Yearbook of Iran*. Tehran: Iran Islamic Republic Statistic Center.

Jupp, V. (2006). *The Sage dictionary of social research methods*: Sage Publications Ltd.

Karami, R., Ismail, M., & Md.Sail, R. (2009). *Gender identity on career aspiration to top management among Iranian graduate students of an institution of higher learning.* Paper presented at the 1st International Conference on Educational Research and Practice, Malaysia.

—. (2011). Gender identity and career aspiration to top management of Malaysian graduate students. *Asian Pacific Journal of Education, 31*(4), 439-454.

Khalil, A. H. O. (2008). *Relationships between extension competencies, organizational commitment and job satisfaction with performance of agricultural extension workers in Yemen (doctoral dissertation).* Doctor of philosophy, University Putra Malaysia, Selangor.

Khalil, A. H. O., Ismail, I., Suandi, T., & Silong, A. D., (2008). Extension worker as a leader to farmers: influence of extension leadership competencies and organizational commitment on extension workers' performance in Yemen. *The Journal of International Social Research, 1*(4), 368-387.

Kline, R. B. (2008). *Promise and pitfalls of structural equation modeling in gifted research. prepublication version submitted April 15, 2008.* Washington, DC: American Psychological Association Books.

—. (2011). *Principles and practice of structural equation modeling*: The Guilford Press.

Kumar, R. (2005). *Research methodology*. London, thousand, Oaks, New Delhi: SAGE Press.

Ladebo, O. J., Olaoye, O. J., & Adamu, C. O. (2008). Extension personnel's self-esteem and workplace relationships: Implications for job satisfaction and affective organizational commitment foci. *The Journal of Agricultural Education and Extension, 14*(3), 249-263.

Ladewig, H., & Rohs, F. R. (2000). Southern extension leadership development: Leadership development for a learning organization. *Journal of Extension, 38*(3).

Landis, R. S., Beal, D. J., & Tesluk, P. E. (2000). A comparison of approaches to forming composite measures in Structural Equation Models. *Organizational Research Methods, 3*(2), 186-207.

Leal, V. M. (2006). *Predictors of non-traditional career self-efficacy in Mexican American adolescent women (doctoral dissertation).* PhD, The Ohio State University USA.

Malhotra, N. K. (2004). *Marketing research an applied orientation* (fourth ed.). USA: Pearson Education International.

Mallinckrodt, B., Abraham, W. T., Wei, M., & Russell, D. W. (2006). Advances in testing the statistical significance of mediation effects. *Journal of Counseling Psychology, 53*(3), 372-378.

Mathieu, J. E., & Taylor, S. R. (2006). Clarifying conditions and decision points for meditational type inferences in organizational behavior. *Journal of Organizational Behavior, 27*, 1031-1056.

Muijs, D. (2004). *Doing Quantitative Research in Education with SPSS.* London, Thousand Oaks, New Delhi: Sage

Neuman, W. L. (2007). *Basics of Social Research Qualitative and Quantitative Approaches* (second ed.). United States of America Pearson Education, Inc.

Powell, G. N., & Butterfield, D. A. (2003). Gender, gender identity, and aspirations to top management. *Women in Management Review, 18*(1/2), 88-96.

Powell, G. N., Butterfield, D. A., & Parent, J. D. (2002). Gender and managerial stereotypes: have the times changed? *Journal of Management, 28* (2), 177-193.

Preacher, K., & Hayes, R. F. (2008). Asymptotic and resampling strategies for assessing and comparing indirect effects in multiple mediator models. *Behavior Research Methods, 40*(3), 879-891.

Ricciardi, D. (2005). *Socializing and supporting new leaders: examining the mentor and Protégé relationship* Paper presented at the Democracy in Educational Leadership: The Unfinished Journey Toward Justice, November Nashville, TN.

Rosenberg, M. (1965). *Society and the adolescent self-image.* Princeton, NJ: Princeton University Press.

Schumacker, R., & Lomax, R. G. (2010). *A beginner's guide to Structural Equation Modeling* (Third ed.). New York: Routledge, Taylor & Francis Group.

Stevens, J. (2002). *Applied multivariate statistics for the social science.* (Fourth edition ed.). Mahwah, New Jersey, London: Lawrence Erlbaum associates, publishers

Trochim, W. K. M. (2006). Probability sampling Retrieved February, 11, 2009, from http://www.socialresearchmethods.net/kb/order.php

Williams, M. (2003). *Making sense of social research* London: SAGE Publication Ltd.

Xu, Q. (2007). *A predictive model of employee self development: the effect of individual and contextual variables (doctoral dissertation).* phd, University of California, Irvine.

CHAPTER SEVEN

ACHIEVEMENT MOTIVATION IN THE LEADERSHIP ROLE OF EXTENSION AGENTS: EVIDENCE

This chapter presents the evidence of factors influencing achievement motivation in leadership role in Iran. The evidence is presented in terms of descriptive statistics and structural equation modelling of achievement motivation in the leadership role of extension agents. The first section begins with the description of the respondents' individual characteristics, followed by background factor variables, personal factor variables, environmental factor variables, and achievement motivation dimensions; the description of the latest construct answered the first objective of the study. The structural equation model was then organized into a direct structural model and full mediating structural model. The direct structural model is a foundation for answering the direct relationships of background, personal, and environmental factors with achievement motivation in the leadership role of extension agents, which is the second objective of the study. The full mediating structural model or overall model covers the third objective of the study and is a basis for answering the mediation effects of personal and environmental factors.

Descriptive Statistics

Descriptive analysis was used to organize and summarize the data (Bahaman & Turiman, 1999). Thus, to summarize and organize the variables with a five point Likert scale in a way that is easier to make sense of and understand, the descriptive analysis is reported in a three point scale, including low (1), moderate (2), and high (3). To do so, first the mean summated score is calculated based on the sum of the items' scores of each scale, divided by the number of the scale's items, resulting in a mean score between 1 and 5. Second, the class interval was calculated by subtracting the highest value (5) from the lowest value (1) of the data set, then the differences were divided by the number of classes and the result was 1.33 (Bahaman & Turiman, 1999).

The third was to determine the location of each case in the three classes of low, moderate, and high. The first class (low level) includes the scores between 1 and 2.33, calculated by adding the lowest value (1) and the class interval 1.33: (1 + 1.33= 2.33). The second class (moderate level) includes the scores between 2.34 and 3.66 resulting from adding 2.33 and the class interval 1.33: (2.33 + 1.33= 3.66). The third class includes the scores between 3.67 and 5.00 that are calculated by adding 3.67 and the class interval 1.33: (3.67 + 1.33= 5.00).

The descriptive statistics include individual characteristics, background factor variables, personal factor variables, environmental factor variables, and achievement motivation.

Individual Characteristics

This section describes a general profile of the respondents' individual characteristics, which consist of gender, educational qualification, current organizational position, and work place location. Table 7-1 shows that around one-fifth (21.8%) of the respondents were female and 78.2% were male. The majority (67.3%) of respondents had a Bachelor of Science degree, while 15.5% of respondents had lower than a Bachelor of Science degree, and 17.2% held an upper level degree. In terms of current organizational position a little more than one-half of the respondents (54.8%) held an expert position and 21.5% held the position of a responsible expert. Thus, 76.3% of respondents experienced different levels of expert positions. 38.9% of respondents worked in the Agricultural Extension and Services Centres, 46.9% worked in Administration of Jihad-e-Agriculture in counties, and 14.2% worked in Organization of Jihad-e-Agriculture in provinces.

Table 7-1: Demographic and professional profile of respondents (n= 303)

Variable	Frequency	Percentage
Gender		
Male	237	78.2
Female	66	21.8
Educational Qualification		
Under diploma	3	1.0
Diploma	26	8.6
Technician	18	5.9
Bachelor	204	67.3
Master	50	16.5
PhD	2	0.7
Current Organizational Position		
Technician	21	6.9
Expert	166	54.8
Responsible expert	65	21.5
First managerial level	40	13.2
Middle managerial level	10	3.3
Top managerial level	1	0.3
Work Place Location		
Centre	118	38.9
County	142	46.9
Province	43	14.2

Background Factor Variables

Age (career stage): Table 7-2 shows that out of 303 respondents, slightly more than half of them (52.5%) were in the establishment of career stage, which, according to Super's theory, is the age between 25 and 44 years old. The next majority group of the respondents (41.6%) was in maintenance of career stage, which is from 45 to 65 years old. The average age of respondents was 41 years old.

Work experience: A cumulative percentage shows around half of the respondents (51.5 %) had work experience less than 15 years, in which the percentage of extension agents with work experience less than 5 years was remarkable (20.5%). The percentage of extension agents with work experience less than 15 years (51.5 %) is consistent with their mean of work experience, which is around 15 years (see Table 3-2).

Gender identity (masculinity and femininity): In order to determine the gender identity of respondents, a total of 30 items were used to determine possession of socially desirable, stereotypically masculine and feminine personality characteristics. Masculine stereotype characteristics were measured with 10 items, feminine stereotype characteristics also with 10 items and 10 filler items. The responses are scored on a seven-point scale from 1 = "Never or almost never true" to 7 = "Always or almost always true". However, they are computed on to a three point scale to summarize the results, in which low ranged from 1 to 3, moderate ranged from 3.1 to 5, and high ranged from 5.1 to 7. Table 7-2 provides the respondents' levels of masculine and feminine gender identity. The majority of the respondents (79.2%) possess a high level of masculinity, similar to the feminine traits, which are also possessed by the majority of respondents to a high level (87.1%). However, there was no respondent possessing a low level of masculine or feminine traits. Only (20.8 %) reported a moderate level of masculinity, and 12.9 % reported a moderate level of feminine traits. Thus, the data indicated that the respondent's average scores of masculinity and femininity possession were 5.62 and 5.90 respectively, which indicates a high level of both traits.

Leadership Stereotype: Assessing the respondent's stereotypical thinking regarding the leadership role, 11 items were developed based on the reviewed literature. Table 7-2 shows the level of the respondents' stereotypical thinking regarding leadership role. The overall mean score was 3.14 and the standard deviation was .92. Half of the respondents (51.2%) scored in the moderate level, while 28.7% and 20.1 of the respondents respectively scored high and low. Thus, the majority of respondents believed in a moderate to high level of stereotypical thinking (79.9%), meaning men have higher ability, better chances, or give priority to the women in leadership achievement.

Table 7-2: Distribution of respondents by background factor variables (n= 303)

Background Factor Variables	Frequency	%	Mean	SD
Age (career stage)			**41.00**	**8.94**
≥24 (exploration)	18	5.9		
25-44 (establishment)	159	52.5		
45-65 (maintenance)	126	41.0		
<65 (disengagement)	-	-		
Work Experience			**15.58**	**9.23**
>5	62	20.5		
6-10	48	15.8		
11-15	46	15.2		
16-20	40	13.2		
21-25	54	17.8		
<26	53	17.5		
Gender Identity (Masculinity)			**5.62**	**0.67**
Low (1.0-3.0)	-	-		
Moderate (3.1-5.0)	63	20.8		
High (5.1-7.0)	240	79.2		
Gender Identity (Femininity)			**5.90**	**0.70**
Low (1.0-3.0)	-	-		
Moderate (3.1-5)	39	12.9		
High (5.1-7.0)	264	87.1		
Leadership Stereotype			**3.14**	**0.92**
Low (1.00-2.33)	61	20.1		
Moderate (2.34-3.66)	155	51.2		
High (3.67-5.00)	87	28.7		

Personal Factor Variables

Self-efficacy: Using the General Self-Efficacy Scale (GSE) designed for the adult population with 10 self-report items, the respondents' sense of perceived self-efficacy was measured. Table 7-3 shows that the majority of respondents (97.7%) perceived moderate (55.1%) to high (43.6%) level of self-efficacy, while only 1.3% stated a low level. The overall mean score was 3.74 with a standard deviation of .65. Thus, the data indicated that almost all of the respondents (97.7%) perceived a moderate to high level of self-efficacy.

Self-esteem: The respondents' self-esteem was measured using 8 items. Table 7-3 provides the respondents' perceived level of self-esteem. Based on the data, more than one-half of the respondents (63.7%) perceived high level of self-esteem. In addition, 36% of respondents reported a moderate level, and a negligible percent (.3) reported low level. The mean of scores (3.91) show respondents mostly perceived a high level of self-esteem.

Achievement value (Subjective, Intrinsic, and Extrinsic): The achievement value as discussed in the review of literature consists of subjective and objective value. The subjective value was assessed using four items, and objective achievement value was measured considering intrinsic and extrinsic achievement values. The intrinsic achievement value using 7 items assessed the personal interests in achieving leadership role for challenging their competencies and skills and making meaningful contributions to society. The extrinsic achievement with 11 items assessed outward personal interests in achieving leadership role, which include receiving more pay, higher position, and advancing at a faster rate. The results, according to Table 7-3, indicated that the majority of respondents (85.2%) chose subjective achievement values at a moderate (43.6%) to high (41.6%) level. Further, the results showed that the majority of respondents reported a moderate level of interest to intrinsic (70.6%) and extrinsic (72.6%) achievement values. Thus, the data indicated that the majority of respondents were moderately interested in objective achievement value (including intrinsic 70.6% and extrinsic 72.6%), while the subjective achievement value was chosen by the majority (85.2%) at a moderate to high level.

Leadership Competency: The level of extension agents' leadership competency was assessed using 18 items. Table 7-3 provides the respondents' perceived level of leadership competency. The findings indicated that the majority (80.5%) of respondents obtained a high score in leadership competency. The overall average score of leadership competency was 4.23, which was more than the mid-point (3) between lowest and highest possible score. The data indicated that the majority of extension agents (80.5%) assessed their leadership competency at a high level.

Table 7-3: Distribution of respondents by personal factor variables (n= 303)

Personal Factor Variables	Frequency	%	Mean	SD
Self-efficacy			**3.74**	**0.65**
Low (1.00 - 2.33)	4	1.3		
Moderate (2.34 - 3.66)	132	43.6		
High (3.67 – 5.00)	167	55.1		
Self-esteem			**3.91**	**0.62**
Low (1.00 - 2.33)	1	0.3		
Moderate (2.34 - 3.66)	109	36.0		
High (3.67 – 5.00)	193	63.7		
Subjective achievement value			**3.43**	**0.94**
Low (1.00 - 2.33)	45	14.9		
Moderate (2.34 - 3.66)	132	43.6		
High (3.67 – 5.00)	126	41.6		
Intrinsic achievement value			**3.23**	**.62**
Low (1.00 - 2.33)	22	7.3		
Moderate (2.34 - 3.66)	214	70.6		
High (3.67 – 5.00)	67	22.1		
Extrinsic achievement value			**3.16**	**0.62**
Low (1.00 - 2.33)	26	8.6		
Moderate (2.34 - 3.66)	220	72.6		
High (3.67 – 5.00)	57	18.8		
Leadership Competency			**4.23**	**0.65**
Low (1.00 - 2.33)	4	1.3		
Moderate (2.34 - 3.66)	55	18.2		
High (3.67 – 5.00)	244	80.5		

Environmental Factor Variables

Mentoring: The mentor support outcomes on career functions were measured using 5 items. Table 7-4 shows one-half of the respondents (51.5%) believed that the mentor support had a high outcome on their career function. A close percentage (43.2%) reported a moderate influence of mentor support on their career function. In general, based on the score average or mean (3.71), the respondents reported higher influence of mentor support on career function than the midpoint average score (3).

Organizational supports: Perceived Organizational Support was assessed using two subscales of organizational work support by 7 items, and organizational non-work support by 5 items. The organizational work support refers to the environment that was desired to promote work goal structure and attainment, along with professionalism, development opportunities, and meritocracy. The organizational non-work support refers to the organizational environment that balances work and family. Table 7-4 shows slightly more than half of the respondents assessed their organizational environment as a moderately supportive in work (58.7%) and non-work (60.1%) matters. Further, a relatively high percentage of respondents (87.7%) evaluated their organization as low (29%) to moderately (58.7%) supportive in work functions, while a cumulative percentage of low (32%) and moderate (60.1%) support in non-work function were at the higher end (92.1%). Thus the average score shows that the organization's work support (2.75) and non-work support (2.65) were nearly the same, and were close to the midpoint average score of 3.

Supervisor supports: The helpful role of extension supervisors is crucial in creating a supportive work and non-work environment. Thus, the supervisor supports, as agents of applying organization strategy, were assessed using two sub-scales of work support with 10 items, and non-work support with 4 items. Both sub-scales show the majority of respondents' supervisors were moderately to highly supportive in work (83.1%) and non-work (85.1%) related subjects. The average score also confirmed that supervisors' work support (3.28) and non-work support (3.23) were nearly the same and slightly exceeded the midpoint average score of 3 (Table 7-4).

Table 7-4: Distribution of respondents by environmental factor variables (n= 303)

Environmental Factor Variables	Frequency	%	Mean	SD
Mentoring			**3.71**	**0.85**
Low (1.00 - 2.33)	16	5.3		
Moderate (2.34 - 3.66)	131	43.2		
High (3.67 – 5.00)	156	51.5		
Organizational work supports			**2.75**	**0.77**
Low (1.00 - 2.33)	88	29.0		
Moderate (2.34 - 3.66)	178	58.7		
High (3.67 – 5.00)	37	12.2		
Organizational non work supports			**2.65**	**0.78**
Low (1.00 - 2.33)	97	32.0		
Moderate (2.34 - 3.66)	182	60.1		
High (3.67 – 5.00)	24	7.9		
Supervisors work supports			**3.28**	**0.91**
Low (1.00 - 2.33)	51	16.8		
Moderate (2.34 - 3.66)	134	44.2		
High (3.67 – 5.00)	118	38.9		
Supervisors non work supports			**3.23**	**0.94**
Low (1.00 - 2.33)	45	14.9		
Moderate (2.34 - 3.66)	148	48.8		
High (3.67 – 5.00)	110	36.3		

Level of Achievement Motivation in the Leadership Role of Extension Agents in Iran

This section under the descriptive statistics answers the first objective of the study. In this study the achievement motivation is defined through three dimensions, including aspiration, mastery, and salience. Aspiration was measured using 9 items and it assessed the respondents' desire to achieve a leadership role. The results, as shown in Table 7-5, indicated a little more than half of the respondents (56.1%) had a high aspiration to achieve a leadership role, followed by 36% with moderate and 24% with low level of aspiration. Thus, 92.1% of the respondents scored moderate to high in terms of aspiration to achieve a leadership role.

Mastery was measured using 8 items to determine the respondents' level of desire for intellectual challenge, and desire to choose difficult, challenging tasks. The finding showed that 49.8% of the respondents scored high, 46.5% scored moderate, and 3.6% scored low. Thus 96.3% of the respondents scored moderate to high in terms of mastery motivation to achieve a leadership role (Table 7-5).

The salience dimension of achievement motivation was measured using 10 items. The scale assesses the extent to which a person sees involvement in leadership role as central to his or her adult life role. The results show that the majority (66.3%) of the respondents scored moderate, followed by 26.4% at high level, and 7.3% at low level.

A comparison of study results about aspiration, mastery and salience showed that around one-half of respondents' aspiration (56.1%) and mastery (49.8%) were measured as high, compared to salience, for which the majority of respondents were at moderate level (66.3%). As a whole, the majority of the respondents' level of motivation to achieve the leadership role in aspiration (92.1%), mastery (96.3%), and salience (92.7%) dimensions of achievement motivation was categorized as moderate to high. This is confirmed by looking at the average score of each dimension of aspiration (3.67), mastery (3.65) and salience (3.27), all slightly above the midpoint of 3 (Table 7-5).

Overall achievement motivation was calculated from summated mean score of three dimensions of aspiration, mastery, and salience. The results, according to Table 7-5, showed that the majority of respondents (60.7%) had a moderate level of achievement motivation, 38.9% had a high level, and only .3% showed a low level of motivation to achieve leadership role. The average score was 3.53 with a standard deviation of .46. Thus, the data provide the answer to the first objective in which the majority of respondents (99.6%) had moderate to high levels of achievement motivation in leadership role.

Table 7-5: Distribution of respondents by achievement motivation dimensions (n= 303)

Achievement Motivation Dimensions	Frequency	%	Mean	SD
Aspiration			**3.67**	**0.77**
Low (1.00 - 2.33)	24	7.9		
Moderate (2.34 - 3.66)	109	36.0		
High (3.67 – 5.00)	170	56.1		
Mastery			**3.65**	**0.63**
Low (1.00 - 2.33)	11	3.6		
Moderate (2.34 - 3.66)	141	46.5		
High (3.67 – 5.00)	151	49.8		
Salience			**3.27**	**0.63**
Low (1.00 - 2.33)	22	7.3		
Moderate (2.34 - 3.66)	201	66.3		
High (3.67 – 5.00)	80	26.4		
Overall achievement motivation			**3.53**	**0.46**
Low (1.00 - 2.33)	1	0.3		
Moderate (2.34 - 3.66)	184	60.7		
High (3.67 – 5.00)	118	38.9		

Structural Equation Modelling of Achievement Motivation in Leadership Role

To answer the second and third inferential objectives of the study and related hypotheses, the structural equation modelling of achievement motivation including direct and overall models were specified and assessed. The structural model was represented by specifying the set of relationships between constructs. The proposed structural model was designed according to Bandura's social cognitive theory and Farmer's model of achievement motivation to explain the relationships between the variables and identify the factors influencing achievement motivation in leadership role. The exogenous variables of this study include the background factor variables (age, gender, feminine gender identity, masculine gender identity, work experience, leadership stereotypes), the personal factor variables (self-efficacy, subjective achievement value, intrinsic achievement value, extrinsic achievement value, extension agent's leadership competency), and the environmental factor variables (mentoring, extension supervisors supports). It is worth saying that gender did not enter into the model due its nominal scale, and its influence investigated through independent-samples t-test, also it was not possible to test the moderating effect of gender as the number of males and females in the sample was not equal or even close. The structural model tested the influence of exogenous variables on the achievement motivation in leadership role as an endogenous variable of the study. Further, the structural model aims to test the mediation effects of the personal and environmental factors variables and achievement motivation in leadership.

In this study, firstly the direct structural model was designed (Figure 7-1) to answer objective two and its related hypotheses. Later, the full mediating structural model, meaning the overall model (Figure 7-2), will be presented to provide the basis for assessment of the third objective in comparison to the direct model.

Direct Structural Model

The assessment of structural models including direct structural models is focused on two issues, comprising of: a) the overall and relative model fit, and b) the size, direction, and significance of the structural parameter estimates, depicted with one-headed arrows on a path diagram (Hair et al., 2006). Table 7-6 shows the result of direct structural model fits, which indicates that the model did fit the data: $\chi2$ (663) = 1177.928, p= .000 $\chi2$/DF=1.777; CFI = .903, IFI= .904, TLI = .892, RMSEA = .051.

Table 7-6: GOF indices of direct structural model

GOF Index	CMIN	CFI	IFI	TLI	RMSEA
Value	1177.928 (P=.000)	.903	.904	.892	.051

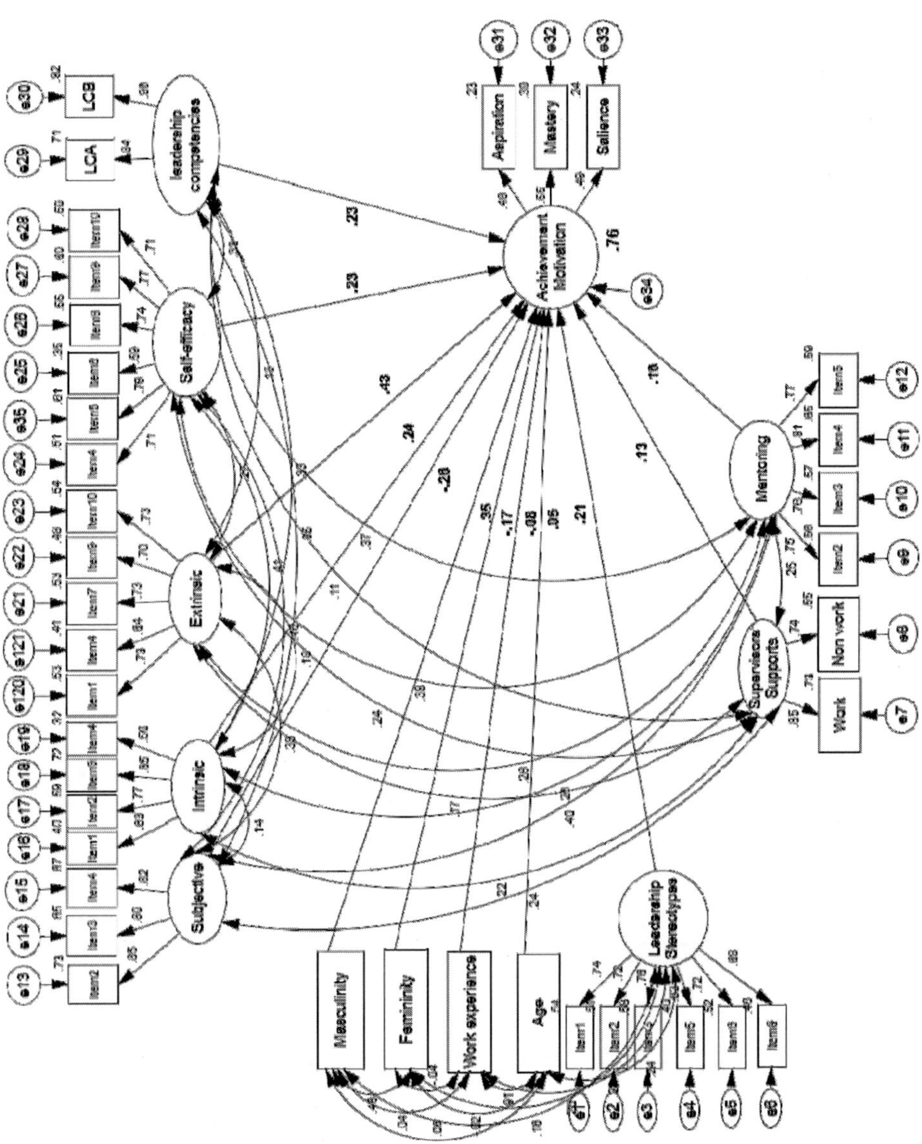

Figure 7-1: The direct structural model of the study

Table 7-7 shows the unstandardized regression weights and standardized regression weights for the hypothesized path in the direct model. The results show that from twelve coefficients associated with the path linking the model's exogenous and endogenous variables, eight were significant by the critical ratio test (> ± 1.96, p< .05). Moreover, the results of the direct structural model revealed that 76% of variance was explained in the proposed direct model of achievement motivation in leadership role by these independent variables (see Figure 7-1). The following section represents the assessment of the second objective based on the direct model.

Table 7-7: Standardized regression weight in direct hypothesized path model

Hypothesized Path			Standardized Regression Weight (β)	P
Masculinity	→	Achievement Motivation	.352	.000
Femininity	→	Achievement Motivation	-.173	.026
Work experience	→	Achievement Motivation	-.076	.637
Age	→	Achievement Motivation	.048	.760
Subjective achievement value	→	Achievement Motivation	-.259	.008
Self-efficacy	→	Achievement Motivation	.228	.017
Leadership stereotypes	→	Achievement Motivation	.207	.009
Mentoring	→	Achievement Motivation	.155	.080
Supervisors supports	→	Achievement Motivation	.128	.113
Leadership competencies	→	Achievement Motivation	.227	.007
Intrinsic achievement value	→	Achievement Motivation	.243	.008
Extrinsic achievement value	→	Achievement Motivation	.429	.000

Background, Personal, and Environmental Factors and Achievement Motivation in the Leadership Role of Extension Agents in Iran

This section aims to answer the second objective of the study and three related hypotheses (H1, H2, H3) as follow:

H1: There is a significant relationship between background related variables (gender, gender identity, work experience, age, and leadership stereotypes,) and achievement motivation of extension agents in leadership role.

The H1 is further divided into sub-hypotheses labelled from H1-1 to H1-5, which are discussed accordingly. To investigate the relationships among the background factor variables in this study the results of the direct model were checked, except for gender, which is a nominal scale variable and its hypothesis is answered using independent-samples t-test.

H1-1: There is a difference among male and female extension agents in terms of achievement motivation in leadership role.

An independent samples t-test was conducted to compare achievement motivation by gender. Table 7-8 shows there was not a significant difference in the achievement motivation for male (M=3.5, SD=.46) and female (M=3.48, SD=0.49) respondents; t (301) =.982, p = 0.327 (2-tailed). The result suggests that gender does not have an influence on achievement motivation in leadership role. Thus, Hypothesis 1-1 was not supported by the study results.

Table 7-8: Results of independent sample t-test on achievement motivation between male and female

Gender group	n	Mean	SD	t	p
Male	237	3.54	.45	.982	.327
Female	66	3.48	.49		

H1-2: There is a difference in the relationship of gender identity (masculinity and femininity) and achievement motivation in the leadership role of extension agents.

Gender identity refers to the sex–appropriate preferences or characteristics an individual develops or learns, rather than characteristics that appear from biological origins. For a while, masculinity and femininity were core dimensions of the human personality. Males highly possess masculine traits, such as independence and dominance. Feminine traits, such as gentleness and sensitivity to the feelings of others, are mostly possessed by females (Powell & Graves, 2003). Feminine and masculine traits could affect achievement motivation of extension agents' leadership role in different ways. Thus, each of the feminine and masculine traits is considered as an observed variable to see their influence on the extension agents' achievement motivation in leadership.

As portrayed in Figure 7-1 and Table 7-7, the exogenous observed variable of masculinity had a significant relationship with achievement motivation in leadership role (β=.352, C.R.=4.253, Sig=.000), whereby for any one unit standardized deviation increase in masculinity, the achievement motivation will be increased by .352. The exogenous observed variable of femininity also had a significant relationship with achievement motivation in leadership role (β=-.173, C.R.=-2.232, Sig=.026). Although for any one unit standardized deviation increase in femininity, achievement motivation will be decreased by .173. Further, the masculinity showed a higher strength (β=.353) according to the estimated Standardized Regression Weights compared to femininity (β=.173). Thus, the H1-2 was supported; this means that there is a difference in the relationship of gender identity (masculinity and femininity) and achievement motivation in leadership role.

H1-3: There is a significant relationship between work experience and achievement motivation in the leadership role of extension agents.

The results of the direct structural model, according to Figure 7-1 and Table 7-7, indicated that there is no significant relationship between work experience and achievement motivation in the leadership role of extension agents (β=-.076, C.R.=-.472, Sig=.637). It is considerable that for any one unit standardized deviation increase in work experience, the achievement motivation in leadership role will decrease by .076. Overall, Hypothesis 1-3 is rejected.

H1-4: There is a significant relationship between age and achievement motivation in the leadership role of extension agents.

As depicted in Figure 7-1 and Table 7-7, the exogenous observed variable of age was found to have no significant relationship with achievement motivation (β=.048, C.R.=.306, Sig=.760). Thus, Hypothesis 1-4 is rejected.

H1-5: There is a significant relationship between leadership stereotypes and achievement motivation in the leadership role of extension agents.

Figure 7-1 and Table 7-7 further show that there is a significant relationship between leadership stereotypes' latent constructs, including six indicators and achievement motivation in the leadership role of extension agents (β=.207, C.R.=2.606, Sig=.009). Therefore, Hypothesis 1-5 was supported.

H2: There is a significant relationship between personal related variables (self-efficacy, self-esteem, achievement value, and extension agents' leadership competencies) and achievement motivation in the leadership role of extension agents.

To investigate the direct effect of personal factor variables on achievement motivation of extension agents in leadership the direct model was tested and to further specify the answers, hypothesis two is divided into sub-hypotheses labelled as H2-1 to H2-4, and each is discussed accordingly.

H2-1: There is a positive relationship between self-efficacy and achievement motivation in the leadership role of extension agents.

According to the results as shown in Figure 7-1 and Table 7-7, there is a significant and positive relationship between the self-efficacy latent construct, including six indicators, and achievement motivation in the leadership role of extension agents (β=.228, C.R.=2.395, Sig=.017).

H2-2: There is a positive relationship between self-esteem and achievement motivation in the leadership role of extension agents.

This hypothesis was not tested since in the initial stage of measurement of the model, the variable was dropped.

H2-3: There is a difference in the relationship of achievement value (intrinsic and extrinsic as objective achievement value and subjective achievement value) and achievement motivation in the leadership role of extension agents.

In this study achievement value is defined through subjective and objective values underlying achievement of leadership role. The objective value refers to the persistence in importance of personal interests in achieving leadership role, including receiving more pay, higher position, advancing at a faster rate (extrinsic values), exercising competencies and skills, and making meaningful contributions to society (intrinsic value). Subjective achievement value refers to the persistence in importance of personal interests, such as work-family balance, in achieving leadership role. Thus, the influence of each achievement value (subjective, extrinsic and intrinsic) on the achievement motivation in the leadership role of extension agents was assessed.

The results as indicated in Figure 7-1 and Table 7-7, show that there is a negative significant relationship between subjective achievement value latent construct, including three indicators, and achievement motivation in the leadership role of extension agents (β=-.259, C.R.=-2.651, Sig=.008). This means that for any one unit standardized division increase in the subjective achievement value, achievement motivation will decrease by .259. The assessment of latent construct of intrinsic achievement value, including four indicators, showed a positive and significant relationship with achievement motivation in the leadership role of extension agents (β=.243, C.R.=2.660, Sig=.008). The latent construct of extrinsic achievement value, with five indicators, also showed a positive and significant relationship with achievement motivation in the leadership role of extension agents (β=.429, C.R.=4.474, Sig=.000). Further, the order of strength among the three mentioned achievement values according to the estimated Standardized Regression Weights is as follows; extrinsic (β=.429), subjective (β=-.259), and intrinsic (β=.243). Thus, H2-3 was supported; this means that there is a difference in the relationship of achievement value (subjective, extrinsic and intrinsic) and achievement motivation.

H2-4: There is a positive relationship between extension agents' leadership competencies and their achievement motivation in leadership role.

The result of the direct structural model (Figure 7-1 and Table 7-7) indicated that there is a significant positive relationship between latent construct of extension agents' leadership competencies with two parcels and achievement motivation in leadership role (β=.227, C.R.=2.694, Sig=.007). Thus, Hypothesis 2-4 was supported.

H3: There is a positive relationship between environment related variables (mentoring, extension organization supports, and extension supervisor supports) and achievement motivation in the leadership role of extension agents.

To investigate the influence of environment factor variables on achievement motivation in leadership of extension agents, the results of the direct model were tested and to further specify the answers the H3 was divided into sub-hypotheses, labelled as H3-1 to H3-3, which are discussed accordingly.

H3-1: There is a positive relationship between mentoring and achievement motivation in the leadership role of extension agents.

As portrayed in Figure 7-1 and Table 7-7, there is no significant relationship between latent construct of mentoring, with four indicators, and achievement motivation in the leadership role of extension agents (β=.155, C.R.=1.751, Sig=.080). Therefore, Hypothesis 3-1 was not supported.

H3-2: There is a positive relationship between extensions organization supports and achievement motivation in the leadership role of extension agents.

This hypothesis was not tested due to the low construct validity of the variable which caused it to be dropped in the initial stage of analysis.

H3-3: There is a positive relationship between extension supervisor supports and achievement motivation in the leadership role of extension agents.

According to the results as shown in Figure 7-1 and Table 7-7, there is no significant relationship between latent construct of extension's supervisor supports, including two parcels of work and non-work support, and achievement motivation in the leadership role of extension agents

(β=.128, C.R.=1.584, Sig=.113). Therefore, Hypothesis 3-3 was not supported.

Full Mediating Structural Model (Overall Model)

The overall structural models were designed to answer the third objective/fourth hypothesis of the study regarding mediation effects of personal and environmental factors on the relationships of background factors and achievement motivation in leadership role of extension agents. The overall structural model, similar to the direct structural model, was assessed through using the GOF indices, and considering the size, direction, and significance of the structural parameter estimates. Table 7-9 shows the result of overall structural model fits, which indicate that the model fit the data: χ^2 (649) = 1170.791, p = .000 χ^2/DF=1.80; GFI = .834, CFI = .902, IFI= .903, TLI = .888, RMSEA = .052.

Table 7-9: GOF indices of overall structural model

GOF Index	CMIN	GFI	CFI	IFI	TLI	RMSEA
Value	1170.791 (P=.000)	.834	.902	.903	.888	.052

Table 7-10 shows the unstandardized regression weights and standardized regression weights for the hypothesized path model, which show that, from forty eight coefficients associated with the path linking the model's exogenous and endogenous variables, twenty three were significant by the critical ratio test (> ± 1.96, p< .05). Moreover, the results of the structural model revealed that 78% of variance was explained in the proposed model of achievement motivation by these independent variables (see Figure 7-2). In summary, the hypothesized achievement motivation in leadership role model was found to fit the data. The next step is to determine the study results in relation to the study hypothesis.

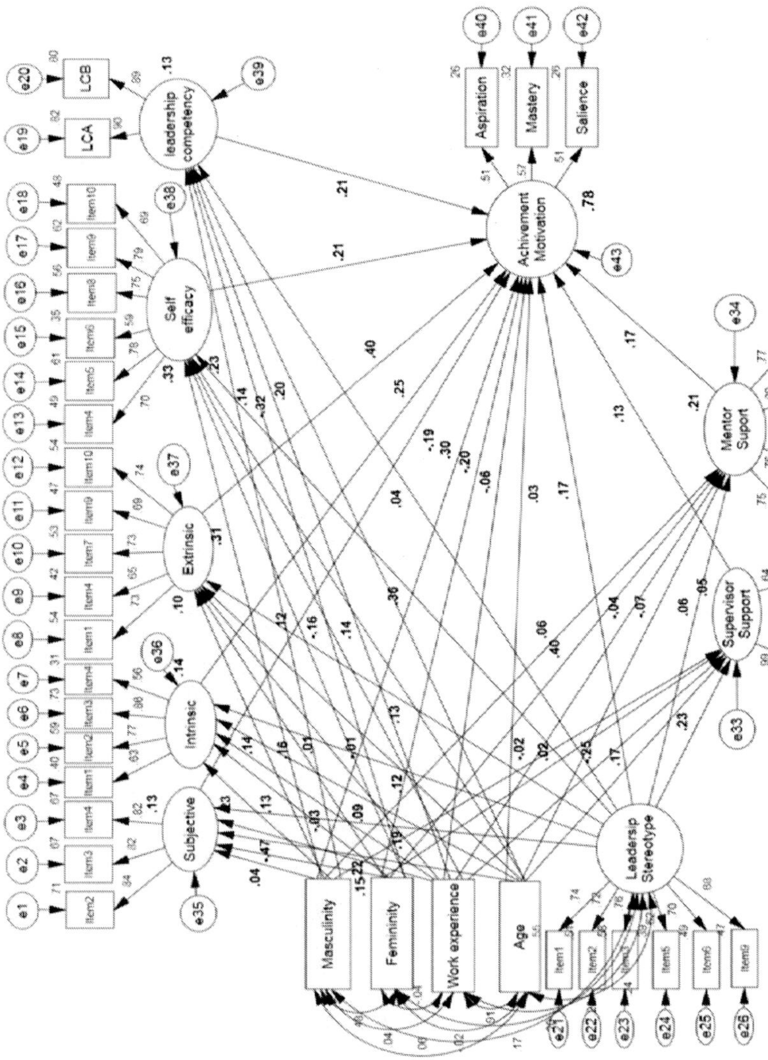

Figure 7-2: The overall structural model of the study showing the relationships between the variables

Table 7-10: Standardized regression weight in overall hypothesized path model

Hypothesized path			Standardized Regression Weights (β)	p
Masculinity	→	Self-efficacy	.313	.000
Masculinity	→	Subjective achievement value	.038	.570
Masculinity	→	Intrinsic achievement value	.226	.001
Masculinity	→	Extrinsic achievement value	.137	.049
Masculinity	→	Leadership competencies	.227	.000
Masculinity	→	Mentoring	.059	.359
Masculinity	→	Supervisors supports	-.019	.771
Femininity	→	Self-efficacy	.119	.053
Femininity	→	Subjective achievement value	.150	.028
Femininity	→	Intrinsic achievement value	.134	.053
Femininity	→	Extrinsic achievement value	.157	.026
Femininity	→	Leadership competencies	.138	.039
Femininity	→	Mentoring	.399	.000
Femininity	→	Supervisors supports	.022	.742
Work experience	→	Self-efficacy	-.161	.218
Work experience	→	Subjective achievement value	-.465	.001
Work experience	→	Intrinsic achievement value	-.029	.845
Work experience	→	Extrinsic achievement value	.013	.929
Work experience	→	Leadership competencies	-.320	.025
Work experience	→	Supervisors supports	-.247	.114
Work experience	→	Mentoring	-.108	.071
Age	→	Self-efficacy	.136	.289
Age	→	Subjective achievement value	.223	.118
Age	→	Intrinsic achievement value	.090	.534
Age	→	Extrinsic achievement value	-.007	.962
Age	→	Leadership competencies	.197	.159
Age	→	Mentoring	-.071	.605
Age	→	Supervisors supports	.174	.237

Hypothesized path			Standardized Regression Weights (β)	p
Leadership stereotypes	→	Self-efficacy	.362	.000
Leadership stereotypes	→	Subjective achievement value	.185	.007
Leadership stereotypes	→	Intrinsic achievement value	.125	.074
Leadership stereotypes	→	Extrinsic achievement value	.131	.066
Leadership stereotypes	→	Leadership competencies	.037	.583
Leadership stereotypes	→	Supervisors supports	.230	.013
Leadership stereotypes	→	Mentoring	.067	.307
Masculinity	→	Achievement Motivation	.303	.000
Femininity	→	Achievement Motivation	-.196	.018
Work experience	→	Achievement Motivation	-.061	.692
Age	→	Achievement Motivation	.031	.827
Leadership stereotypes	→	Achievement Motivation	.171	.049
Self-efficacy	→	Achievement Motivation	.213	.014
Subjective achievement value	→	Achievement Motivation	-.193	.009
Intrinsic achievement value	→	Achievement Motivation	.247	.002
Extrinsic achievement value	→	Achievement Motivation	.397	.000
Leadership competencies	→	Achievement Motivation	.208	.005
Mentoring	→	Achievement Motivation	.171	.029
Supervisors supports	→	Achievement Motivation	.129	.048

Mediation Effect of Personal and Environmental Factors on the Relationships between Background Factor Variables and Achievement Motivation in the Leadership Role of Extension Agents in Iran

This section aims to answer the third objective of the study and related hypothesis (H4) as follow:

H4: Personal and environmental factors mediate the relationship between background factor variables and achievement motivation in the leadership role of extension agents.

To test the mediation effects of personal and environmental factor variables on the relationship between exogenous variables and endogenous variables, first the normal methods of mediation testing (Baron & Kenny, 1986) cited in Mallinckrodt et al. (2006), were used to compare the results of the overall and direct structural model for each variable (Table 7-7, 7-10). Therefore, the mediation effect of each variable of background factors was discussed in a specific hypothesis (H4-1 to H4-5) as in the following section.

H4-1: Personal and environmental factors mediate the relationship between masculinity and achievement motivation in the leadership role of extension agents.

Based on Baron and Kenny's approach, the results of testing the mediation effects on the relationships between masculinity and achievement motivation were tested through four conditions. The first condition, which is the significant relationship between masculinity and achievement motivation, as mentioned in Table 7-7, is met (β=.352, Sig=.000). The second condition is to investigate the significant relationship between masculinity and mediator variables (personal factor variables including: self-efficacy, subjective, intrinsic, extrinsic, leadership competency; and environmental variables including: mentoring and supervisors supports). The results of overall the model, as shown in Table 7-10 and Figure 7-2, indicated that masculinity had no significant relationship with mediator variables namely; subjective (β=.038, Sig=.570), mentoring (β=.059, Sig=.359), and supervisors supports (β=-.019, Sig=.771). Thus, the hypothesis of mediation effects for these three variables was not supported. The significant indirect effect of masculinity on the rest of the mediator variables, including self-efficacy (β=.313, Sig

=.000), intrinsic (β=.226, Sig=.001), extrinsic (β=.137, Sig=.049), and leadership competency (β=.227, Sig =.000) showed signs of possible mediation effects in which we need to test a third step for their support or rejection (Table 7-10).

The third condition was to see the significant indirect effect of possible mediator variables (see the result of second condition) on achievement motivation. The results indicated in Table 7-10 show that all of the possible mediator variables resulting from the second step had a significant relationship with achievement motivation including self-efficacy (β=.213, Sig=.014), intrinsic (β=.247, Sig=.002), extrinsic (β=.397, Sig=.000), and leadership competency (β=.208, Sig=.005). Then, we need to investigate the last step or condition to see whether the mediation effect of the four mentioned variables will support and at which level. Thus, the results of the direct (Table 7-7) and overall (Table 7-10) structural model were compared to test the effect of masculinity on the achievement motivation through direct or mediating variables. The reduced amount of Standardized Regression Weight that remains significant in this relationship from the direct model (β=.352, Sig=.000) when compared to the overall model (β=.303, Sig=.000) supported the partial mediation effects of self-efficacy, intrinsic, extrinsic, and leadership competency in the relationship between masculinity and achievement motivation.

H4-2: Personal and environmental factors mediate the relationship between femininity and achievement motivation in the leadership role of extension agents.

According to Baron and Kenny's approach, the first condition of testing mediation effect was to test the significant direct relationship between femininity and achievement motivation as mentioned in Table 7-7, in which the condition was met (β=-.173, Sig=.026). The second condition is to investigate the significant relationship between femininity and mediator variables (personal factor variables including: self-efficacy, subjective, intrinsic, extrinsic, leadership competencies; and environmental variables including: mentoring and supervisors supports). The results of indirect effect in the overall model, as shown in Table 7-10 and Figure 7-2, indicated that femininity had no significant relationship with mediator variables, namely self-efficacy (β=.119, Sig=.053), intrinsic (β=.134, Sig=.053), and supervisor supports (β=.022, Sig=.742). Thus, the hypothesis of mediation effects for these three variables was not supported. The significant indirect effect of femininity on the rest of the mediator variables, including subjective (β=.150, Sig=.028), extrinsic (β=.157, Sig=.026), leadership competencies (β=.138, Sig=.039), and mentoring (β=.399, Sig=.000), showed signs of possible mediation effect, for which we need to test a third step to decide on its support or rejection (Table 7-10).

The third condition was to see the significant indirect effect of possible mediator variables (see the result of second condition) on the achievement motivation. The results according to the Table 7-10 show that all of the possible mediator variables resulting from the second step show a significant relationship with achievement, including mentoring (β=.171, Sig=.029), subjective (β=-.193, Sig=.009), extrinsic (β=.397, Sig=.000), and leadership competencies (β=.208, Sig=.005). Thus, we need to investigate the last step or condition to see whether the mediation effect of the four mentioned variables will be supported or not. The results of direct (Table 7-7) and overall (Table 7-10) structural models were compared to test the effect of femininity on the achievement motivation through the mediating variables. The Standardized Regression Weight of the direct model (β=-.173, Sig=.026) remains significant in the overall model (β=-.196, Sig=.018) supporting the partial mediation effects of subjective, extrinsic, and leadership competency in the relationship between femininity and achievement motivation.

H4-3: Personal and environmental factors mediate the relationship between work experience and achievement motivation in the leadership role of extension agents.

The first condition of Baron and Kenny's approach, which is the significant relationship between work experience and achievement motivation, as mentioned in Table 7-7, was not met (β=-.076, Sig=.637). But according to Kenny et al. (1998) cited in Mallinckrodt et al. (2006), the first condition of Baron and Kenny's approach is not necessary. Thus, the next condition was tested as explained below.

The second condition was to investigate the significant relationship between work experience and mediator variables (personal factor variables including: self-efficacy, subjective, intrinsic, extrinsic, leadership competencies; and environmental variables including: mentoring and supervisors supports). The results of indirect effect in the overall model, as shown in Table 7-10 and Figure 7-2, indicated that work experience had no significant relationship with mediator variables, namely self-efficacy (β=.161, Sig =.218), intrinsic (β=-.029, Sig=.845), extrinsic (β=.013, Sig=.929), and mentoring (β=-.247, Sig=.114), and supervisor supports (β=.108, Sig=.071). The significant indirect effect of work experience on the subjective (β=-.465, Sig=.001), and leadership competencies (β=-.320, Sig=.025), showed signs of possible indirect effects in which we need to test a third step (Table 7-10).

The third condition was to see the significant indirect effect of subjective and leadership competencies on the achievement motivation. The results, according to Table 7-10, show that there is a significant relationship between subjective achievement value (β=-.193, Sig=.009) and achievement motivation, similar to the leadership competency (β=.208, Sig=.005) and achievement motivation. Thus we could conclude that there is indirect effect for work experience → subjective → achievement motivation, similar to the, work experience → leadership competency → achievement motivation.

H4-4: Personal and environmental factors mediate the relationship between age and achievement motivation in the leadership role of extension agents.

The standardized direct effect of age on achievement motivation as shown in Table 7-7 was not significant (β=.048, Sig=.760). Also, none of the mediator variables (Table 7-10) show a significant relationship with age, including self-efficacy (β=.136, Sig=.289), subjective (β=.223, Sig=.118), intrinsic (β=.090, Sig=.534), extrinsic (β=-.007, Sig =.962), leadership competencies (β =.197, Sig=.159), and mentoring (β=-.071, Sig=.605), and supervisor supports (β=.174, Sig=.237). As a result, there was no reason to test the rest of the conditions. Therefore, *H4-4* was rejected, since age had no significant relationship with achievement motivation or mediator variables.

H4-5: Personal and environmental factors mediate the relationship between leadership stereotype and achievement motivation in the leadership role of extension agents.

Table 7-7 shows the significant direct effect between leadership stereotypes and achievement motivation (β=.207, Sig=.009) which is the first condition in testing the mediation effect. The second condition is to investigate the significant relationship between leadership stereotype and mediator variables (personal factor variables including: self-efficacy, subjective, intrinsic, extrinsic, leadership competencies; and environmental variables including: mentoring and supervisors supports). Table 7-10, which is the result of indirect effect in the overall model, indicated that leadership stereotype had no significant relationship with mediator variables, namely intrinsic (β=.125, Sig=.074), extrinsic (β =.131, Sig=.066), leadership competency (β=.037, Sig =.583), mentoring (β=.037, Sig =.583) and supervisor support (β =.067, Sig =.307). Thus, the hypothesis of mediation effect for these five variables was not supported. The significant indirect effect of leadership stereotype on the rest of the mediator variables including self-efficacy (β=.362, Sig = .000), subjective (β=.185, Sig=.007), and mentoring (β =.230, Sig=.013) showed signs of possible mediation effect for which we need to test the third step (Table 7-10).

The third condition was to see the significant indirect effect of possible mediator variables (see the result of second condition) on the achievement motivation. Table 7-10 shows that all of the possible mediator variables resulting from the second step had a significant relationship with achievement, including self-efficacy (β=.213, Sig=.014), subjective (β=-.193, Sig=.009), and mentoring (β =.171, Sig=.029). Thus, we need to investigate the last step or condition to see whether the mediation effect of the four mentioned variables will be supported or not, and if it is supported the level of support needs to be considered. Then, the results of direct (Table 7-7) and overall (Table 7-10) structural models are compared to test the influence of leadership stereotype on the achievement motivation through the mediating variables. The reduced amount of Standardized Regression Weight remains significant in this relationship from the direct model (β= .207, Sig=.009) in comparison with the overall model (β=.171, Sig=.049), which supported the partial mediation effects of self-efficacy, subjectivity, and mentoring in the relationship between leadership stereotype and achievement motivation in leadership role.

Further, the bootstrap method was used as a recent way of testing mediation effects (Table 7-11). This method allows one to see the overall mediating effect of all mediators' variables on the relationship between exogenous variables and achievement motivation in leadership role. To perform the bootstrap analysis, we used the graphical interface of Amos 18 software package that directly produces bootstrapped bias-corrected confidence intervals for indirect effects. We requested 5000 bootstrap samples (typically it is at least 1000, but some authors such as Hayes (2009) recommended at least 5000), drawn by default with replacement from the full data set of 303 cases. Also, requested were 95% confidence intervals for the bias-corrected.

Table 7-11: The effect of exogenous variables on achievement motivation through the mediation of personal and environmental factors

Variables	Point Estimate	S.E.	Bootstrapping		
			BC Percentile 95% CI		
			Lower	Upper	Two Tailed Sig. (BC)
Masculinity	.224	.058	.117	.348	.000
Femininity	.191	.060	.082	.319	.001
Work experience	-.063	.108	-.285	.135	.464
Age	.069	.099	-.114	.274	.392
Leadership stereotype	.173	.065	.057	.323	.004

Note: BC = Bias- corrected confidence interval; 5,000 bootstrap samples was requested.

The study results using bootstrapped bias-corrected confidence intervals with the requested 5000 bootstrap samples, and 95% confidence intervals, as indicated in Table 7-11, show that personal and environmental factors significantly mediate the relationship between masculinity (CI=.117 to .348, Sig=.000), femininity (CI=.082 to .319, Sig=.001), leadership stereotype (CI=.057 to .323, Sig=.004), and achievement motivation in leadership role. The results could not support the mediation effects of personal and environmental factors on the relationship between work experience (CI=-.285 to .135, Sig=.464), age (CI=-.114 to .274, Sig=.392), and achievement motivation in leadership role. This means the results of the bootstrap method confirmed the hypotheses H4-1, H4-2, H4-5, while the results of the mediation effect for H4-3, H4-4 were not supported.

Conclusion

This chapter comprises the evidence that was planned to contribute to a better understanding of the factors influencing achievement motivation in leadership role of extension agents in Iran. The results of the study are presented in descriptive and structural equation modelling of achievement motivation in the leadership role, in responding to the objectives and hypotheses of the study. The study results regarding background factors support the differences of gender identity and achievement motivation, in addition to significant relationships of leadership stereotype and achievement motivation; while the influence of gender, work experience, and age were not supported. The personal factors that include intrinsic, extrinsic, self-efficacy, and leadership competencies, showed a positive direct relationship with achievement motivation, while subjective achievement value negatively affected the achievement motivation in leadership role. The direct relationships of environmental factors, including mentoring and supervisor supports, were not supported. The relationships of masculinity and achievement motivation were partially mediated through self-efficacy, intrinsic, extrinsic, and leadership competencies. Also the study results supported the partial mediation effects of mentoring, subjective, extrinsic, and leadership competencies in the relationship between femininity and achievement motivation. There were indirect effects for work experience through subjective achievement value and leadership competency on achievement motivation in leadership role. Lastly, the results supported the partial mediation effects of self-efficacy, subjectivity, and mentoring in the relationship between leadership stereotype and achievement motivation in leadership role.

References

Bahaman, A. S., & Turiman, S. (1999). *Statistics for social research with computer application*. Malaysia: University Putra Malaysia Press.

Baron, R. M., & Kenny, D. A. (1986). The moderator-mediator variable distinction in social psychological research: conceptual, strategic, and statistical considerations. *Journal of Personality and Social Psychology, 51*(6), 1173-1182.

Byrne, B. M. (2010). *Structural equation modeling with AMOS: basic concepts, applications, and programming* (second ed.). New York, London: Taylor and Francis Group.

Coffman, D. L., & MacCallum, R. C. (2005). Using parcels to convert path analysis models into latent variable models. *Multivariate Behavioral Research, 40*(2), 235-259.

Garson, D. (2009). Structural Equation Modeling. Retrieved 1 December, 2009, from http://faculty.chass.ncsu.edu/garson/PA765/structur.htm

Hair, J. F., Black, W. C., Babin, B. J., & Anderson, R. E. (2010). *Multivariate data analysis*. United States of America: Pearson prentice hall.

Hair, J. F., Black, W. C., Babin, B. J., Anderson, R. E., & Tatham, R. L. (2006). *Multivariate data analysis* (sixth ed.). United States of America: Pearson prentice hall.

Hayes, A. F. (2009). Beyond Baron and Kenny: statistical mediation analysis in the new millennium. *Communication Monographs, 76*(4), 408-420.

Kline, R. B. (2011). *Principles and practice of structural equation modeling*: The Guilford Press.

Mallinckrodt, B., Abraham, W. T., Wei, M., & Russell, D. W. (2006). Advances in testing the statistical significance of mediation effects. *Journal of Counseling Psychology, 53*(3), 372-378.

Powell, G. N., & Graves, L. M. (2003). *Women and men in management.* Thousand Oaks, London, New Delhi: SAGE Publications.

Schumacker, R., & Lomax, R. G. (2010). *A beginner's guide to Structural Equation Modeling* (Third ed.). New York: Routledge, Taylor & Francis Group.

Schumacker, R. E., & Lomax, R. G. (2004). *A beginner's guide to Structural Equation Modeling* (Second ed.). Mahwah, New Jersey, London: Lawrence Erlbaum Associates.

CHAPTER EIGHT

ACHIEVEMENT MOTIVATION IN THE LEADERSHIP ROLE OF EXTENSION AGENTS: FUTURE AGENDA

This book has illustrated and examined three main factors influencing achievement motivation in the leadership role of extension agents. Bandura's Social Cognitive theory (1989; 1999; 2006) and Farmer's (1985) model of career achievement motivation were the foundation of establishing factors on which the book and the hypotheses of study in Iran were formulated. Bandura's Social Cognitive theory, that counts the influence of personal and environmental factors on human behaviour, had not been approached or tested under the pre-influence of background factors. Background factors more specifically in leadership role, as a traditionally masculine role, are of vital importance. Although Farmer's (1985) model made claims on the influence of background factors, it did not approach job role achievement motivation in the workplace and related variables, which is a new addition in the present book. Another originality of the present book is considering leadership role achievement motivation with three dimensions of aspiration, mastery and salience, which is consistent with Farmer's model. Although aforementioned originalities of the present book and research cause some limitations in founding the literature to support the findings, replacing it with the closest literature was attempted. Thus, this final chapter aims to discuss the interpretation of the findings regarding the factors influencing achievement motivation and the mediating effect of personal and environmental factors. Further, this chapter presents interpretation of study findings compared with past research, conclusions, and future agenda in relation to achievement motivation in the leadership role of extension agents in Iran.

Background, Personal, and Environmental Factors and Achievement Motivation

This section provides the interpretation of study findings compared with past research to show how the findings fit in with pervious works, and to justify the probable inconsistencies. In fact, the opposite results are not always a matter of research internal validity, although it could be true, but it is a venture into the unknown as there is no ultimate "should be".

Achievement motivation: In this study, the majority of respondents (60.7%) reported a moderate level of achievement motivation. However, there are variations in the past study results regarding extension agents' leadership role attainment in which Fabusoro et al. (2008) reported low status of leadership attainment among extension agents in Nigeria, while Khalil, Ismail, Suandi, and Silong (2009) reported a moderate level in Yemen. Most importantly, little is known about the level of motivation to attainment of leadership role among extension agents in Iran. Sadighi (2006), in studying Iran's extension agents, reported a low general motivation level of extension agents in performing their overall job. However, the findings of the current study do not support Sadighi's (2006) results, which could be related to the more specific view of this study on achievement motivation in leadership role in comparison to investigation of overall job motivation.

Background Factor: The study results using independent-samples t-test show there is no significant difference among male (M=3.5, SD=.46) and female (M=3.48, SD=0.49) extension agents and achievement motivation in leadership role. Ismail (2008) explained that the common beliefs of sex differences are the origin of gender difference ideology in the workplace. Actually, gender differences, as Powell and Graves (2003) revealed, symbolize beliefs about male and female roles in the community and workplace. Also gender differences were manifested in stereotypes, prejudice, and discrimination, which mostly are applied to women. It could be argued that gender does not cause differences, while stereotypical assumptions cause gender differences, such as we found about gender identity and leadership stereotype. Further, the study results showing no significant difference among men and women extension agents in leadership achievement motivation are consistent with Karami et al. (2009) in the study of Iranian post graduate students in a Malaysian university that gender did not influence significantly respondents' aspiration to achieve top management. Similarly, Lopes (2006), in studying the differences in dimensions of career motivation between international and domestic graduate students at Pennsylvania State University, supported these findings of no significant relationship between gender and achievement motivation. Further, the study of Akiyoshi (2007), in investigating the effects of tertiary education on Japanese women's employment status and career aspirations, showed gender did not relate to career aspiration. Again this is consistent with Borgen and Betz (2008) in a univariate analyses that indicated there were no significant gender differences on leadership, motivation, or academic achievement.

However, there are some contradictions, such as the work of Cannon (2004), Farmer (1985), and London and Noe (1997), whose studies showed gender was significantly related to all three measures of career identity. The possible explanation for the contradictory results according to O'Neill and O'Reilly (2004) is the importance of gender identity rather than sex (a biological construct) as determinant of career success or achievement. They argued that what defines an individual is the combination of sex and gender identity and when a researcher investigates males' and females' differences through only sex, or only gender identity, conflicting findings emerge.

Research findings identified a significant relationship between masculinity and achievement motivation (β=.352, C.R.=4.253, Sig=.000) in which, for any one unit of standardized deviation increase in masculinity, the achievement motivation would be increased by .352. Also, femininity had a significant relationship with achievement motivation (β=-.173, C.R.=-2.32, Sig=.026) in which, for any one unit of standardized deviation increase in the femininity, achievement motivation would be decreased by .173. Further, masculinity showed a higher strength (β=.352) according to the estimated Standardized Regression Weights in comparison with femininity (β=.173). This study produced results which corroborate the findings of the previous work in this field. Karami et al. (2009) measured gender identity using BSRI, which showed that gender identity explained a large proportion of career aspiration variance, that femininity influences people's career aspiration, and brings a lower level of career aspiration, which is a dimension of achievement motivation. Also the results of logistic regression analysis of Karami et al. (2009) are similar to those of Powell and Butterfield (2003) using a similar measurement, which found masculinity had a significant positive effect on aspiration to top management (Powell and Butterfield, 2003; Karami et al., 2009).

The results of this study did not show significant relationship between work experience (β=-.076, C.R.=-.472, Sig=.637), age (β=.048, C.R.=.306, Sig=.760) and achievement motivation of extension agents in leadership role. The findings of the current study do not support previous research, such as that of Dorsen, Carlson and Goodyear (2006) about the effect of work experiences on one dimension of achievement motivation among students. The differences of the current results from existing literature could be explained from three possible aspects. First, the non-significant relationships of work experience and achievement motivation in leadership role may be related to the low self-complexity of respondents in which they could not accumulate experience in leadership roles, which produced such a result. According to Zoogah and Abbey (2010), self-complexity is related to the number of self-aspects, such as roles, relationships, contexts, or activities, and the amount of individual independence concerning those self-aspects. They asserted that it is self-complexity of the employee that determines the amount of experience they accumulate from previous roles, activities, and relationships in a variety of contexts. Individuals with high self-complexity tend to accumulate more experiences than those with low self-complexity, and these experiences may be leveraged to enhance individual performance (Zoogah & Abbey, 2010), one of which is achievement motivation in leadership role. Second, as explained in the context of the study in Iran, the first three decades of extension jobs were more about transfer of technology than human resource development, and the use of leadership role in extension is a recent move in both academia and practice. Thus, it is predictable that work experience in the transfer of technology and seeing clients as passive, helpless reactors who have no knowledge and should learn in a top-down process, did not lead to the accumulation of experience in leadership role. Hence, the study found no association between work experience and achievement motivation in leadership role. Third, the study results that support the influence of work experience on achievement motivation are from other cultures and there was almost no similar study found in the Iranian cultural situation. This could be a possible reason for the study result differences. In sum, it is an area that requires further research.

Further, this study has been unable to demonstrate that age influences achievement motivation, as hypothesized, while Lopes (2006) indicated some relationships between age and the three dimensions of career achievement motivation. However, researchers such as Lopes (2006) and Ottino (2009) asserted that there are mixed results regarding the relationships between age and achievement motivation. In the study of London (1993), older staff showed as much career motivation as younger staff and some elements of career motivation increased with age. While, according to Lange, Yperen, Heijden, Bal, and Dahan (2010), older workers relative to younger workers, showed a significantly higher amount of dominant mastery-avoidance goals that bring the lowest work engagement in social and personal meanings of work. This difference from present study results can be explained in part by counting the effect of similarity in the respondents' age, in which nearly half of them (52.5%) were in the establishment career stage with ages between 25 and 44 years.

Leadership stereotypes were found to affect achievement motivation (β=.207, C.R.=2.606, Sig=.009) showing stereotypical belief regarding male priority in leadership roles, in a population with a majority of males, positively increases achievement motivation in leadership role. Although in reviewing the literature, no data was found on the association between leadership stereotype and achievement motivation, this finding corroborates the ideas of Bandura, who asserted occupation stereotypes would affect personal attributes and behaviour (Leal, 2006).

Personal Factor: The current study found a significant and positive relationship between self-efficacy (β=.228, C.R.=2.395, Sig=.017) and achievement motivation in leadership role among extension agents, almost all of whom (97.7%) perceived a moderate to high level of self-efficacy. This is consistent with the results of Leal (2006) who also found that aspirations and mastery motivation are positively related to non-traditional career self-efficacy. Consistently, Larson et al. (2007) found that self-efficacy significantly contributes to students' choice in program majors and in their future career goals such as career aspirations (or achievement motivation). The finding shows that, in order for the extension agents to be motivated to achieve a leadership role, possession of self-efficacy is a positive variable in which extension agents who assess themselves as being capable of carrying out a leadership role are more motivated to achieve a leadership role.

The study results also indicate a positive relationship between intrinsic achievement value (β=.243, C.R.=2.660, Sig=.008), extrinsic achievement value (β=.429, C.R.=4.474, Sig=.000), and achievement motivation, while the significant relationship between subjective achievement value and achievement motivation of extension agents in leadership role (β=-.259, C.R.=-2.651, Sig=.008) was negative. Rayan and Deci (2000) believed that people have different amounts and different kinds of motivation that affect their level of motivation and the orientation of that motivation. Orientation of motivation involves the fundamental attitudes and goals that provide persistence for an achievement, such as the hypothesis orientation of the present research as objective and subjective. The influence of intrinsic objective value is supported by the study of Farmer (1985) that found significant relationships between the intrinsic achievement value and achievement motivation. Although the existent positive relationship between extrinsic achievement value and achievement motivation at first glance was contradictory with the studies of Vansteenkiste et al. (2007), which asserted that an extrinsically motivated employee has extra negative job outcomes rather than positive job outcomes, the difference can be explained in part due to the type of motivation to achieve; it does not matter what the outcome is to the organization. This means that extension agents are motivated to achieve a leadership role because of something distinct from the job itself, such as having influence over others. The effect of subjective achievement value of the present research could be supported further by Farmer's (1985) explanation regarding negative effects of competing activities on achievement motivation, stemming from family and home interests on the salience dimension of achievement motivation.

The study result shows a positive and significant relationship between leadership competencies (β=.227, C.R.=2.694, Sig=.007) and achievement motivation in the leadership role of extension agents, the majority of whom (80.5%) assessed their leadership competency at a high level. This study produced results which corroborate the findings of Scheer, Ferrari, Earnest, and Connors (2006), who presented a model for preparing extension professionals. They illustrated a model with the competencies required for success in the extension profession, one of which is leadership. Further, the present study was partly consistent with Khalil et al. (2009) who found that extension agents' competencies had a positive and significant relationship with extension worker performance (r=.621; p<.006). Also, leadership competencies, which are the application of knowledge, technical skills, and personal characteristics, lead to outstanding performance (Boyd, 2004). Thus, having leadership competency, which is conceptualized in this study as the knowledge, abilities, and skills used by extension agents to guide the clients, was found as a reason motivating extension agents to achieve a leadership role. This implies that leadership achievement requires related competencies that need to be considered in the academic curriculum of extension courses, as well as on the job training contents for extension agents.

Environmental Factor: The study results indicated that mentoring (β=.155, C.R.=1.751, Sig=.080) and supervisor supports (β=.128, C.R.=1. 584, Sig=.113) had no significant direct relationships with achievement. However, Day and Allen's (2004) study results found mentored individuals did report higher levels of job motivation than those who had not been mentored, which is contradictory with the present study results; but perhaps in this study mentoring mostly is affected through variables other than direct effect. The non-significant relationships of extension supervisor supports and achievement motivation is different from other study results that claimed higher perceived supervisor supports for work and family, and reported greater organizational commitment (O'Neill et al., 2009), which is a dimension of achievement motivation. Also, according to Powell and Graves (2003), having a family supportive supervisor reduces work-family conflict and gives employees more encouragement to stay with the organization and motivates them to achieve job roles in particular and higher performance in general. A possible explanation for the inconsistency of the present research results and reviewed literature may be related first to the stronger mediation effects of these variables than the direct affect that past studies did not consider. Second, the lack of adequate perceived support also may make the relationships non-significant, while the descriptive analysis results, which were based on the average score of supervisors work support (3.28) and non-work support (3.23), show a moderate level of perceived support.

Mediation Effect of Personal and Environmental Factors

The study results about the mediation effect of personal factor variables in the relations of masculine (self-efficacy, intrinsic and extrinsic achievement value, and leadership competency) and feminine (mentoring, subjective, extrinsic, and leadership competencies) gender identity on achievement motivation are consistent with Powell and Graves' (2003) explanation. Powell and Graves (2003) introduced individual predominant gender role as a variable affecting an individual's achievement value and personal attributes, and consequently their motivation to achieve. The individuals with different gender identities use different types of evaluation value in assessing career achievement. Masculine characters most likely preferred objective measures and feminine characters are the reverse.

Further, leadership is a role in which its achievement could satisfy different needs of different individuals with different orientations. The needs for sympathy and sensitivity to others are some of the common characteristics of leadership roles, which are consistent with feminine traits and related to subjective achievement value. The need for dominance, strong personality and leadership ability are some other common characteristics of leadership roles, which are consistent with masculine traits and could affect the individual orientation to objective value, such as intrinsic and extrinsic achievement value (Powell and Graves, 2003). It is consistent with the present research findings that the effects of masculinity on achievement motivation were mediated through leadership competency, intrinsic and extrinsic achievement value and bring higher achievement motivation; while femininity mediates through subjective achievement value that later brings lower achievement motivation. However, the findings of the current study regarding mediation effects of extrinsic achievement value on the relationships of femininity and achievement motivation differs from the previous research, which mostly discussed the subjective value of feminine traits. However the mediation effects of mentoring are in agreement with feminine traits in which support is desired.

The mediation effects of self-efficacy on the relationships between masculinity and achievement motivation are supported by phenomena such as human motivation and accomplishment (Pajares & Urdan, 2006), well-being, and success in life (Burke, 2005), in which the phenomena could not be explained without discussing the function of self-efficacy. Bandura asserted that perceived efficacy does not only directly affect people's behaviour, but also indirectly affects the other determinants, such as the behaviour people decide to follow, the challenging tasks they choose to do, how much effort they put forth in given tasks, as well as their commitment to the tasks (Bandura, 2006). Thus, one could conclude that self-efficacy, in addition to the direct effect, has mediation effects on the relationships of an individual's character and all three dimensions of achievement motivation: aspiration, mastery and salience. The abovementioned literature, along with the study results of Judge et al. (2007), confirmed that individual differences (such as trait orientation) could affect the predictive validity of self-efficacy and achievement motivation.

The study results using bootstrapping could not support the mediation effects of personal and environmental factors on the relationship between work experience (CI=-.285 to .135, Sig= .464), age (CI=-.114 to .274, Sig= .392), and achievement motivation. However, the results have been unable to demonstrate that work experience is mediated through personal and environmental factors. But work experience was indirectly related with achievement motivation through leadership competency and subjective achievement value. Thus, the study results confirm the idea of social cognitive theory about the effect of work experience on cognitive ability (Mills et al., 2007) and consequently achievement motivation. But this is inconsistent with the study results of Dorsen, Carlson and Goodyear (2006) on work experiences and career aspiration of students, since the present study found signs of a negative indirect effect of work experience through subjective and leadership competency on achievement motivation. Mills et al. (2007) indicated that individuals through the process of reflective self-examination, estimate the sufficiency of their thoughts, behaviour, and motivation, and modify their thinking, and later their behaviour. One possible explanation is that the extension agents, through the working years, experienced the process of reflective self-examination, but did not see leadership competency as an important matter in their work. They also paid attention to caring about their work and life balance, as in subjective achievement value. Possibly the new generation of extension agents are trained and possess a higher level of leadership competency than the experienced extension agents, due to the recent emphasis on extension agents' leadership role noted by Vijayaragavan, Singh, and Wason (2005), which is a possible reason for the signs of a negative effect of work experience on achievement motivation.

However, London and Noe (1997) found that personality characteristics related with achievement motivation altered at the same time as an individual aged because of adaptation to the environment. But the present study results could not show direct or mediation effects on the relationship of age and achievement motivation. Also, contradictory to Greller (2000), who revealed that in different age groups, norms dictated a different career motivation, the present study results did not show such a norm dictation for different age groups of extension agents.

The study results using bootstrapping indicated that personal and environmental factors significantly mediate the relationship between leadership stereotype (CI=.057 to .323, Sig=.004), and achievement motivation in leadership role. The results contribute to the idea of Bandura (1978) cited in Leal (2006), that occupational stereotypes would affect the personal attributes in general and self-efficacy in particular, especially for females, for whom it consequently affects achievement motivation. Despite the extensive increase in the percentage of women in leadership roles over the past three decades, and more emphasis on feminine traits in leadership, the idea still persists that a good manager is a person who mainly possesses masculine traits (Powell et al., 2002). Also, consistent with the present research results about the mediation effects of self-efficacy on the relation of leadership stereotype and achievement motivation, Leal (2006) partially supported the idea that occupational stereotypes have important effects on work self-efficacy.

Conclusion

The study results provide evidence in support of a theoretical framework illustrated in the first part of the book. The proposed framework in the overall model of the study predicted 75% of variability in leadership achievement motivation, which is an acceptable percentage to conclude that the presented framework is an evidence based framework for future research. Furthermore, it is necessary to re-emphasize the three main factors influencing achievement motivation in leadership role by having in mind their interrelations with each other. Thus, the three following sections present the conclusions based on the background, personal, and environmental factor influences on achievement motivation.

Background Factor Influences

Background factors were included in the research from primary dimensions of diversity, which are not easily changeable or unchangeable, and are not isolated from other dimensions that play an important role in a person's achievement motivation. Researchers often ignore the interdependence of these with other (secondary) dimensions of diversity, which are changeable personal characteristics that are acquired and may be modified or abandoned throughout life. Although secondary dimensions of diversity (personal factors) in comparison with the primary dimensions (background factor) are more controllable, they will be affected by other people's decisions, which in this research we see as environmental factors. Thus, having in mind the prevalence of stereotypical thinking about leadership, the background factor variables that shape gender quality, such as gender role stereotypes, which in the workplace are followed by leadership stereotypes along with career stage and work experience, are considered. The importance of gender quality is related to the low gender ratio of female extension workers and the importance of gender homophily of extension workers with clients due to cultural norms in which female extension clients should be approached by female extension workers. The conclusions regarding background factors are as follows:

First, the study results on significant direct and indirect relationships of masculine and feminine traits of gender identity and leadership stereotype showed the contribution of these variables in explaining achievement motivation in the leadership role of extension agents. More specifically, gender identity and leadership stereotype, according to the bootstrap analysis, were recognized as gender quality variables that could play a primary role in shaping extension agents' personal and environmental factors toward achievement motivation of leadership roles. Recognizing gender identity and leadership stereotype as influential in motivating extension agents to achieve leadership roles are important considerations in planning a new extension environment; one in which, despite a low gender ratio of female extension agents, women should not perceive themselves as unable or inferior in relation to achieving leadership role because of the effect of perceived gender identity and stereotypical assumptions on taking leadership role, and not women's low ability in achieving leadership role. Therefore, it could be concluded that by controlling and understanding the effect of gender identity and leadership stereotype, leadership roles would be achievable for men and women to the same degree. This would be consistent with the study results using a t-test, stating that there are no significant differences between male and female extension agents toward achievement motivation in leadership role. Further, in extension environments where women clients need to be taught and led by the same gender leader, the verified variables could help women extension agents to be motivated to achieve a leadership role; as could knowing that leadership is not a gender specified attribute, and can be acquired by men or women.

Second, a majority of extension agents believed from a moderate up to a high level of stereotypical thinking, which means men have higher ability, better chance, and are given more priority than women in leadership role achievement. From this expression it could be concluded that leadership is still a masculine role. Further, the positive relationship of masculinity and achievement motivation of leadership role confirmed that possession of masculine traits could motivate the extension agents to achieve a leadership role, while the opposite is true for feminine traits. It is a critical issue which may limit women, since women mostly possess feminine traits, which showed a negative relationship with achievement motivation. This issue needs to be considered in extension context, as women extension agents are required to play leadership roles among women clients. Thus, the extension agents must move from stereotypical thinking toward a balanced situation in which leadership roles, according to the latest literature, require both feminine and masculine traits. The extension organizations are also responsible for providing learning and developmental opportunity for extension agents to know more about correcting the stereotypical assumptions of leadership role, which according to the literature, the correction on the stereotypical thinking requires constant negotiation.

Third, extension agents with masculine gender identity reflect a higher intention to objective achievement value (including intrinsic and extrinsic), which is positively associated with achievement motivation in leadership role. Also, possession of masculine traits affects extension agents' belief about their capability to perform leadership roles (self-efficacy) and leadership competencies, which both positively influence achievement motivation in leadership roles.

Fourth, extension agents who possess feminine traits have more intention to subjective achievement value, which is negatively associated with achievement motivation of leadership roles. Further, it could be concluded that the environmental situation of the extension organization was not supportive to extension agents with subjective achievement value, since the individuals with subjective value are motivated to achieve leadership role if the environmental support coexists. Subjective value refers to caring about the family, individual interest, and work balance, which negatively affects achievement motivation in leadership roles. Also, feminine extension agents are more likely to desire mentoring that could support them to stay motivated for achieving leadership role.

Fifth, the extension agents' stereotypical beliefs about leadership roles are a source for self-efficacy establishment that influences achievement motivation in leadership roles. Furthermore, leadership stereotype, through subjective achievement value and mentoring, was associated with achievement motivation in leadership role.

Sixth, work experience of extension agents indirectly, through subjective achievement value and leadership competencies, was associated with achievement motivation in leadership role.

As a whole, it is concluded that in a stereotypically masculine role such as leadership, gender identity and leadership stereotype play vital roles as primary dimensions of diversity in the background factors that, through secondary dimensions of diversity such as personal factors, as well as environmental factors, influence achievement motivation in leadership role.

Personal Factor Influences

The personal factors refer to perception of extension agents regarding a person's cognitive process, such as belief about his or her ability, which acts as an important mediator of motivation and influences as expectations for future performance. Actions taken were measured through self-efficacy, achievement value, and leadership competencies. The following conclusions were made based on the study results:

First, possessing self-efficacy and leadership competency could help and intensify the extension agents in achieving their leadership role. Self-efficacy and leadership competency are two important personal variables for extension agents' empowerment and consequently extension organization performance. It is also expected that empowerment of extension agents in leadership role could in turn empower extension clients to be able to realize their own problems and probable solutions, and consequently become less dependent on extension agents.

Second, for extension agents to be motivated in leadership roles, they need to have leadership competency, such as gentleness and sensitivity to the feelings of others, which are associated with feminine traits, and independence, which is more associated with masculine traits. It is related to the finding that leadership competencies mediate the relationships of femininity and masculinity on achievement motivation that is consistent with the recent theories of leadership. The theories have placed somewhat more emphasis on feminine than masculine characteristics, and femininity has emerged as a needed managerial trait that is gaining acceptance in today's workplace.

Third, partial mediation effect of self-efficacy on the relationships of masculinity and leadership stereotype on achievement motivation is of importance due to knowing that the perceived gender identity and leadership stereotype could cultivate specific self-efficacy affecting extension agents' motivation to achieve a leadership role. Thus, extension organization investment on reducing the stereotypical assumption about leadership role or personal traits could raise extension agents' beliefs about their capability to successfully perform a leadership role (self-efficacy) and later raise their motivation to achieve a leadership role.

Fourth, extension agents' intrinsic and extrinsic achievement values positively contribute toward achievement motivation in leadership role, while the subjective achievement value negatively influences achievement motivation in leadership role.

Fifth, achievement value dimensions were found to be important mediators of masculine and feminine traits. It was found that objectively motivated masculine extension agents show a positive relationship with achievement of leadership role, while feminine traits bring subjective achievement value and only motivate extension agents to achieve leadership role if organizational support is available to them, which in this study was negative.

It is concluded that all variables of personal factors, including self-efficacy, achievement value, leadership competency, in addition to having direct influences on achievement motivation, partially mediate the relationships between different background variables and achievement motivation in leadership role. Thus, the results confirm the mediating role of personal factors, as Bandura theorized, in the relationships of background factors and achievement motivation behaviour.

Environmental Factor Influences

The environmental factor variables in this study include organization and supervisor supports, in addition to mentoring. Considering men's and women's needs and incentives, a supportive environment that leads to achievement of excellence in work-related behaviour was considered as a situation in which both work and family supports were provided. Thus, by looking at the foundation of environmental factor variables and the study results, the conclusions are made as follows:

First, extension agents perceived a low to moderate level of organizational work and non-work support, but low construct validity of this variable, perhaps due to the respondent bias, caused it to be dropped from the inferential analysis. As a result, the researcher is unable to draw a conclusion about the relationships of this variable with other proposed variables. The extension agents assessed their supervisors' as moderate to highly supportive in both work and non-work support, but relationships of extension supervisor supports and achievement motivation were not supported.

Second, extension agents perceived a high to moderate level of mentoring and they believed that mentoring partially mediated the relationships of femininity, leadership stereotype, and achievement motivation in leadership role.

Future Agenda

This section presents the theoretical and practical implications of the book and the study in accordance with the study results and conclusions.

Theoretical Implications: The most important theoretical contributions of this book are on the conceptual framework design that improves our knowledge about achievement motivation in the workplace, with a main focus on emergent leadership roles, since there is a theoretical gap about achievement motivation in the workplace, particularly on emergent leadership roles. Further, the study is a substantiation of Bandura's social cognitive theory, incorporating the pre-influence of background factors on personal and environmental factors, which most past studies did not consider. The study results are an extension to Farmer's model that was designed to predict students' career expectations, while the present framework improves the knowledge of achievement motivation in the workplace and more particularly on emergent leadership achievement motivation in extension context. Therefore, this study contributes to the application of Bandura's social cognitive theory and Farmer's model of achievement motivation in the investigation of achievement motivation of emergent leadership roles. Specifically, this study is the first of its kind to look at the pre-influence of background factors and mediation effects of personal and environmental factors in the extension system context in Iran.

The achievement motivation concepts in this study were conceptualized in three dimensions according to Farmers' point of view, while past studies mostly defined and measured achievement motivation through one dimension. The study is even unique in its focus on expanding the achievement value in terms of objective and subjective achievement values. While past research worked on the extrinsic and intrinsic achievement values, which are categorized as objective achievement values, the importance of subjective achievement value considered in this research is neglected. The subjective achievement value is mostly important to the individual with feminine characteristics, who seeks a balance between work and family. Further, by considering the desire of feminine character individuals, the organizational supports and extension supervisor supports were investigated from work and non-work support aspects. The other contribution of this research to the body of knowledge is in designing and developing the leadership competencies scale for extension agents based on their unique functions in emergent leadership role.

Practical Implications: Achievement motivation provides a strong foundation for a good life. Individuals who are motivated to achieve are dynamic and feel in control of their life. Thus, they pursue goals they have identified for themselves. Achievement motivated people are the ones who make things happen and get results, and this extends not only to getting results through their organizations, but achieving goals in their community (Kerr, 2009). Thus, understanding factors motivating achievement of leadership roles is useful in the leadership development programs for Iran's agricultural extension policy makers, extension organizations, extension agents, and extension clients in particular, and the world in general. Understanding factors influencing achievement motivation in leadership role assists in improving the capability of extension agents to perform their job effectively. Decentralization of leadership structure and delegating more responsibility to the employee will create a greater sense of ownership of the projects, which are characterized by high performance, trust (Duncan et al., 2008), and satisfaction. The empowerment of extension agents in leadership roles, in addition to increasing personal achievement, would increase extension organization achievement by providing higher productivity, efficiency, and satisfaction. This is because delegating more responsibilities to the extension agents through motivating them to achieve a leadership role creates a greater sense of ownership and commitment to the organization. Consequently, there will be higher performance and satisfaction. Further, as times and conditions change, motivational strategies used in the past become less effective, and agricultural managers and policy makers require newer and more reliable empirical information for better decision making (Fabusoro et al., 2008). Therefore, understanding factors influencing achievement motivation in leadership roles is useful for agricultural extension policy makers and organizational managers in the formulation of human resource development policy and strategies in extension systems.

When extension agents are said to be empowered by possessing a leadership role, then they will in turn empower their clients to achieve high efficiency and productivity of their enterprise that consequently enables clients to become less dependent on extension agents in their own problem solving process. Leadership competencies of extension clients are a criterion for cooperative success, which are introduced as the best-suited institutions to develop agriculture in developing countries (Androulidakis, 2001). Thus, the study results would assist leadership development of rural people, since extension agents act as the role models to the clients in terms of leadership competencies and functions.

The fact that background factors influence achievement motivation in leadership roles could be capitalized on in the future learning of extension agents. One way to do this is to consider gender quality variables for decreasing the negative effects of stereotypical assumption and providing a balanced and improved situation. The importance of gender quality becomes one of the concerns of the workplace in general and the extension system in particular, hence, gender ratio and gender homophily of extension workers with clients should be seriously considered in the extension systems due to cultural norms in which women extension clients should be approached by women extension workers. The pressures of stereotypical thinking, according to Luk-Fong & Brennan (2010), are more complex than the dichotomy of femininity and masculinity and have not gone away, needing constant negotiation. Thus, the educational policy implications of this study in extension context can be explored this regard to how negotiation and support are achieved to change the stereotypical assumptions about leadership role.

The results of direct effect of personal factors on achievement motivation, along with their mediation effects in leadership role, could assist extension organizations to develop appropriate training programs. The training programs on leadership competencies, self-efficacy, and achievement values are important due to their significant direct and indirect effects on achievement motivation. Further, the results have implications for organization selection and recruitment, as self-efficacy, achievement value, and leadership competencies become criteria to be used in the recruitment process if extension organizations require or expect their extension agents to engage in effective leadership roles. A kind of measurement scale for these variables could be developed for selection and recruitment purposes.

The study results showed that environmental factor variables have partial applied implications for nurturing and developing extension agents in organizations. Organizational and supervisory supports, along with mentoring, may increase motivation in leadership role achievement. It is not only the work related support which brings higher achievement motivation, but also non-work supports are vital for the extension agents who strongly possess subjective achievement value. Thus, the need for providing the appropriate strategy and teaching process to prepare work support and family friendly culture (non-work support) in extension organizations is one of the important practical implications of this research. Last but not least, the development of mentoring programs can help extension agents in career functions, which provides opportunities to learn new skills and understand the unwritten rules, as well as assist in meeting the deadlines of given tasks, and suggesting specific strategies for achieving career goals. At the same time, extension organizations could see the potential of mentoring as a powerful learning and developmental strategy, which could be used in extension training without taking an employee out of the job or paying lots of money.

Recommendations for Future Research

The most important recommendation of this book is to apply the pre-influence of background factors in gender investigation of future studies, instead of considering the moderation role for gender. Gender does not only affect the strength of relationships, but, through variables such as gender identity, could also play a pre-influencing role on a person's cognitive abilities and perceived environmental situation. Further, it is gender identity of a person that affects their situations, not biological sex. Thus, it is highly suggested that future studies which investigate gender differences among employees in a particular setting, apply both gender and gender identity.

A limitation of the study presented in the second part of the book was to stay away from testing reciprocal relationships of personal and environmental factors and their effect on achievement motivation in leadership roles, due to complexity of framework and needed analysis, which is suggested for future research. Perhaps using a smaller number of variables in personal factors could assist future researchers to cover the reciprocal determinism.

In order to be able to generalize the research framework beyond leadership, future research could address other roles of extension agents by slightly modifying the research framework. Researchers could also study the framework in other countries to compare the results among developing and developed countries or even to test the model in contexts other than extension organizations to see the determinants of achievement motivation in leadership roles.

A conflicting result of this study was the negative influence (although not significant) of work experience on achievement motivation, which needs a deeper study to see how experience in the work place could result in lower achievement motivation. While the instrument used in this study for organizational support and self-esteem was an established and validated scale, it is a self-report instrument open to respondent bias. Future studies should attempt to avoid this in order to provide a clearer picture of the relationship between organizational supports, self-esteem, and achievement motivation.

References

Akiyoshi, M. (2007). Women are women?: The effects of tertiary education on Japanese women's employment status and career aspirations, *The Annual Meeting of American Sociological Association.*

Androulidakis, S. (2001). Leadership ability dimensions in relation to rural women's personal characteristics: A methodological model for further consideration *Journal of International Agricultural and Extension Education 8*(1), 39-46.

Bandura, A. (1989). Social cognitive theory. *Annals of child development, 6*, 1-60.

—. (1999). Social cognitive theory: An agentic perspective. *Asian Journal of Social Psychology 2*, 21-41.

—. (2006). Guide for constructing a self-efficacy scales. In F. Pajares, & Urdan, T. C. (Ed.), *Self-efficacy beliefs of adolescents.* United State of America: IAP-Information Age Publishing, Inc.

Borgen, F. H., & Betz, N. E. (2008). Career self-efficacy and personality: linking career confidence and the healthy personality. *Journal of Career Assessment, 16*(1), 22–43.

Boyd, B. L. (2004). Extension agents as administrators of volunteers: competencies needed for the future. *Journal of Extension, 42*(2).

Burke, M. U. (2005). *Family influence on self-efficacy: exploring the relationship between perceived parenting style, current social support, and self-efficacy belief in a sample of nontraditional college students (doctoral dissertation)* PhD, Capella University, Minnesota.

Cannon, S. C. (2004). *Supporting first-year college women in math, science, and related majors: A Career Development Intervention (Doctoral dissertation).* Doctor of philosophy, North Carolina State University, United States.

Day, R., & Allen, T. D. (2004). The relationship between career motivation and self-efficacy with protégé career success. *Journal of Vocational Behavior, 64*, 72-91.

Dorsen, J., Carlson, B., & Goodyear, L. (2006). Connecting informal STEM experiences to career choices: identifying the pathway. A literature review produced by the ITEST Learning Resource Center. Retrieved 29, Jan, 2009, from http: www.itestlrc.edc.org/connecting-informal-stem-experiences-career

Duncan, D. W., Navarro, M., & Mykerezi, P. K. (2008). *Learning about sustainability: the contribution of the global seminar educational model.* Paper presented at the 24nd Annual Conference of AIAEE "Global Entrepreneurship in International Agricultural and Extension Education", March 9 to15, Costa Rica

Fabusoro, E., Awotunde, J. A., Sodiya, C. I., & Alarima, C. I. (2008). Status of job motivation and job performance of field level extension agents in Ogun State: Implications for agricultural development. *Journal of Agricultural Education and Extension, 14*(2), 139-152.

Farmer, H. S. (1985). Model of career and achievement motivation for women and men. *Journal of Counseling Psychology, 32*(3), 363-390.

Greller, M. M. (2000). Age norms and career motivation. *The International Journal of Aging and Human Development 50*(3), 215 - 226

Judge, T., Jackson, C., Shaw, J., Scott, B., & Rich, B. (2007). Self-efficacy and work-related performance: the integral role of individual differences. *The Journal of Applied Psychology 92*(1), 107-127.

Karami, R., Ismail, M., & Md.Sail, R. (2009). *Gender identity on career aspiration to top management among Iranian graduate students of an institution of higher learning.* Paper presented at the 1st International Conference on Educational Research and Practice, Malaysia.

Kerr, B. (2009). *Encyclopedia of giftedness, creativity, and talent.* USA: SAGE publication, Inc.

Khalil, A. H. O., Ismail, M., Suandi, T., & Silong, A. D. (2009). Human resource development competencies as predictors of agricultural extension agents' performance in Yemen. *Human resource development international, 12*(4), 429-447.

Lange, A. H. d., Yperen, N. W. V., Heijden, B. I. J. M. V. d., Bal, P. M., & Dahan, H. M. (2010). Dominant achievement goals of older workers and their relationship with motivation-related outcomes. *Journal of Vocational Behavior, 77*, 118-125.

Larson, M. L., Wei, M., Wu, T. F., Borgen, H. F., & Bailey, C. D. (2007). Discriminating among educational majors and career aspirations in Taiwanese undergraduates: the contribution of personality and self-efficacy. *Journal of Counseling Psychology, 54*(4), 395-408.

Leal, V. M. (2006). *Predictors of non-traditional career self-efficacy in Mexican American adolescent women (doctoral dissertation).* PhD, The Ohio State University USA.

London, M. (1993). Career motivation of full- and part-time workers in mid and late career. *The International Journal of Career Management, 5*, 21-29.

London, M., & Noe, R. A. (1997). London's career motivation theory: An update on measurement and research. *Journal of Career Assessment, 5*(1), 61-80.

Lopes, T. (2006). *Differences in dimensions of career motivation between international and domestic graduate students by age and gender (doctoral dissertation).* Doctor of Philosophy, The Pennsylvania State University United States.

Ismail, M. (2008). *Gender & career: realities and challenges.* Inaugural lecture series at UPM, Serdang: University Putra Malaysia Press.

Mills, N., Pajares, F., & Herron, C. (2007). Self-efficacy of college intermediate French students: relation to achievement and motivation. *Language Learning 57*(3), 417–442.

O'Neill, A. O., & O'Reilly, A. C. (2004). *Women's careers: the impact of sex and gender identity on career attainment.* Research paper No. 1775(R). Graduate School of Business; Stanford University.

O'Neill, J. W., Harrison, M. M., Cleveland, J., Almeida, D., Stawski, R., & Crouter, A. C. (2009). Work–family climate, organizational commitment, and turnover: Multilevel contagion effects of leaders. *Journal of Vocational Behavior, 74*, 18-29.

Ottino, K. L. (2009). *Diminished aspiration: women central office administrators and the superintendency (doctoral dissertation).* PhD, The University of Minnesota United State.

Pajares, F., & Urdan, T. C. (2006). *Self-efficacy beliefs of adolescents.* United State of America: IAP-Infornation Age Publishing, Inc.

Powell, G. N., & Butterfield, D. A. (2003). Gender, gender identity, and aspirations to top management. *Women in Management Review, 18*(1/2), 88-96.

Powell, G. N., Butterfield, D. A., & Parent, J. D. (2002). Gender and managerial stereotypes: have the times changed? *Journal of Management, 28* (2), 177-193.

Powell, G. N., & Graves, L. M. (2003). *Women and men in management.* Thousand Oaks, London, New Delhi: SAGE Publications.

Ryan, R. M., & Deci, E. L. (2000). Intrinsic and extrinsic motivations: Classic definitions and new directions. *Contemporary Educational Psychology, 25*( ), 54-67.

Sadighi, H. (2006). *Factors affecting agricultural extension personnel's motivation level.* Paper presented at the 22nd Annual Conference of AIAEE "International Teamwork in Agricultural and Extension Education", May14 to19 Clearwater beach, Florida.

Scheer, S. D., Ferrari, T. M., Earnest, G. W., & Connors, J. J. (2006). Preparing extension professionals: The Ohio State University's model of extension education. *Journal of Extension, 44*(4).

Vansteenkiste, M., Neyrinck, B., Niemiec, C. P., Soenens, B., Witte, H. D., & Broeck, A. V. d. (2007). On the relations among work value orientations, psychological need satisfaction and job outcomes: A self-determination theory approach. *Journal of Occupational and Organizational Psychology, 80*, 251-277.

Vijayaragavan, K., Singh, P., & Wason, M. (2005). *Developing training modules for improving management skills of extension professionals.* Paper presented at the 21st Annual Conference" Educational, Extension and Research Strategies for a Changing World", San Antonio, TX, pp. 13-24.

Zoogah, D. B., & Abbey, A. (2010). Cross-cultural experience, strategic motivation and employer hiring preference: An exploratory study in an emerging economy. *International Journal of Cross Cultural Management, 10*(3), 321-343.